Linan, Amable.

Fundamental aspects
 of combustion.

BAKER & TAYLOR BOOKS

Fundamental Aspects of Combustion

THE OXFORD ENGINEERING SCIENCE SERIES

1. D. R. RHODES: *Synthesis of planar antenna sources*
2. L. C. WOODS: *Thermodynamics of fluid systems*
3. R. N. FRANKLIN: *Plasma phenomena in gas discharges*
4. G. A. BIRD: *Molecular gas dynamics*
5. H.-G. UNGER: *Planar optical waveguides and fibres*
6. J. HAYMAN: *Equilibrium of shell structures*
7. K. H. HUNT: *Kinematic geometry of mechanisms*
8. D. S. JONES: *Methods in electromagnetic wave propagation* (Two volumes)
9. W. D. MARSH: *Economics of electric utility power generation*
10. P. HAGEDORN: *Non-linear oscillators* (Second edition)
11. R. HILL: *Mathematical theory of plasticity*
12. D. J. DAWE: *Matrix and finite element displacement analysis of structures*
13. N. W. MURRAY: *Introduction to the theory of thin-walled structures*
14. R. I. TANNER: *Engineering theology*
15. M. F. KANNINEN and C. H. POPELAR: *Advanced fracture mechanics*
16. R. H. T. BATES and M. J. McDONNELL: *Image restoration and reconstruction*
17. K. HUSEYIN: *Multiple-parameter stability theory and its application*
18. R. N. BRACEWELL: *The Hartley transform*
19. J. WESSON: *Tokamaks*
20. P. B. WHALLEY: *Boiling, condensation, and gas-liquid flow*
21. C. SAMSON, M. Le BORGNE, and B. ESPIAU: *Robot control: the task function approach*
22. H. J. RAMM: *Fluid dynamics for the study of transonic flow*
23. R. R. A. SYMS: *Practical volume holography*
24. W. D. McCOMB: *The physics of fluid turbulence*
25. Z. P. BAZANT and L. CEDOLIN: *Stability of structures: elastic, inelastic, fracture, and damage theories*
26. J. D. THORNTON: *Science and practice of liquid-liquid extraction* (Two volumes)
27. J. VAN BLADEL: *Singular electromagnetic fields and sources*
28. M. O. TOKHI and R. R. LEITCH: *Active noise control*
29. I. V. LINDELL: *Methods for electromagnetic field analysis*
30. J. A. C. KENTFIELD: *Nonsteady, one-dimensional, internal, compressible flows*
31. W. F. HOSFORD: *Mechanics of crystals and polycrystals*
32. G. S. H. LOCK: *The tubular thermosyphon: variations on a theme*
33. A. LIÑÁN and F. A. WILLIAMS: *Fundamental aspects of combustion*

Fundamental Aspects of Combustion

Amable Liñán
Department of Motor Propulsion and Fluid Dynamics
School of Aeronautical Engineering
City University
Madrid, Spain

Forman A. Williams
Center for Energy and Combustion Research
Department of Applied Mechanics and Engineering Sciences
University of California, San Diego
La Jolla, California

New York   Oxford
OXFORD UNIVERSITY PRESS
1993

Oxford University Press

Oxford New York Toronto
Delhi Bombay Calcutta Madras Karachi
Kuala Lumpur Singapore Hong Kong Tokyo
Nairobi Dar es Salamm Cape Town
Melbourne Auckland Madrid

and associated companies in
Berlin Ibadan

Published by Oxford University Press, Inc.,
200 Madison Avenue, New York, New York 10016

Oxford is a registered trademark of Oxford University Press

Library of Congress Cataloging-in-Publication Data

Liñán, Amable,
Fundamental aspects of combustion/Amable Liñán, Forman A. Williams.
p. cm. (The Oxford engineering science series; 33)
Includes bibliographical references and index.
ISBN 0-19-507626-5
1. Combustion. I. Williams, F. A. (Forman Arthur), 1934– .
II. Title. III. Series.
QD516.L48 1993 92-23050
541.3′61–dc20

9 8 7 6 5 4 3 2 1

Printed in the United States of America
on acid-free paper

TO Y. B. ZEL'DOVICH

*Who Erected the Foundation
of the Modern Science of Combustion*

FOREWORD

The history of mankind has been intertwined with combustion from antiquity and will remain so in the future. This monograph focuses on the science of combustion rather than its technological, societal, or philosophical aspects. A systematic overview of the science is presented, and current topics of research in the field are addressed, with suggestions and predictions for future development. The intent is to provide the flavor of the subject and to attract researchers from other disciplines.

PREFACE

About five years ago Amable Liñán invited Forman Williams to present a lecture series on the science of combustion at the Instituto de España in Madrid. A stipulation was that lecture notes were to be written and translated into Spanish for publication. Although draft versions of notes were indeed written when the lectures were presented in November of 1987, pressures of other commitments prevented their completion at that time, despite all good intentions. Now, through the encouragement of Oxford University Press, we have revised and updated those notes to prepare the present volume.

The book is addressed to an audience of mature and scientifically literate readers, in both industry and academia, who have not specialized in combustion. It is intended to outline the subject for them in a systematic way and to exhibit to them its mathematical character and its physical and chemical content. In a further effort to impart some of the excitement that we share for the subject, we also include some historical observations, discussions of the technological utility of combustion, identifications of current research problems, and speculations on future developments. The resulting breadth of perspective notwithstanding, the presentation penetrates quickly to the depths of the subject, pulling no punches—there seems no other way to get at the essence of so many combustion problems in so few pages. Despite our decision to address knowledgeable readers, there still is not enough space for us to include everything that interests us in combustion science. In its technical content, in many respects this monograph is largely a condensation of some of the more salient information in the much longer book of Williams, *Combustion Theory* (Addison-Wesley, 1985), which itself is not light reading. However, because of its broader-brush treatment, the present volume will be easier to digest. It could serve as a supplementary text in graduate courses on combustion and as a guide to future research pursuits for specialists in the field, even though it is not intended to be addressed principally to students or researchers in combustion.

The scientific meat extends from the third section of the first chapter through the fifth chapter. We adopt premixed flames and diffusion flames as the main subdivisions of the subject (Chapters 2 and 3) and emphasize both the important roles played by asymptotic analysis and the diversity arising from real chemical kinetics in flames. The somewhat different subjects of ignition, flammability, and detonation are all covered in a single chapter (Chapter 4), in an attempt to emphasize their common features. Turbulent combustion, in both premixed and nonpremixed systems, is an important topic spanning such a wide variety of aspects that an entire chapter (Chapter 5) is devoted to it. It is our opinion that the topics in these few chapters comprise the core of the science of combustion. Many of the trimmings (propellant combustion, combustion instabilities, spray combustion, etc.) are encountered peripherally, in passing, in our presentation, but the primary focus is intended to remain on the core of the science. Although the references to the literature certainly are not thorough, bibliographies are given at the end of the chapters, largely comprised of books and review articles, in an effort to direct the reader to the recent research publications.

Our collaboration in research on the science of combustion began in the early 1960s. For thirty years now, we have continually shared the occasional joy of discovery and the unending excitement of endeavoring to increase our understanding of combustion. Although we both completed our graduate education at Cal Tech, we were not there at the same time and did not meet until later, when we recognized through writing our strongly common research interest. We feel especially fortunate to have lived in an age in which international collaboration of the kind that we have enjoyed is possible. Our friendship has become one of our most precious possessions. We hope that this short volume can help to bring to others our fascination with our chosen science.

We are indebted to the Instituto de España for initiating the activity that resulted in this monograph. We also owe gratitude to various government funding agencies, notably AFOSR, NSF, NASA, ARO, and DOE, for supporting our combustion research over the years. Sincere appreciation is extended to our typist-editor, Sheri Lindelsee, and to our artist-illustrator, Ted Velasquez; without Sheri, in particular, this volume would not have been completed.

Madrid   A. L.
La Jolla, Calif.   F. A. W.

CONTENTS

1. Background and Formulation of Combustion 3

History of the Technology and Science of Combustion 3
Applications 5
Conservation Equations 7
Classifications of Combustion Processes 13
Deflagrations and Detonations 17
Bibliography 20

2. Premixed Flames 21

Simplified Description of Flame Structure 22
Asymptotic Analysis of Flame Structure 28
Flame Instabilities 32
Influences of Kinetic Mechanisms 38
Examples of Real Flames 45
Aspects of Deflagrations in Need of Further Investigation 55
Bibliography 55

3. Diffusion Flames 57

Applications of Diffusion Flames 57
Structures of Laminar Diffusion Flames 61
Diffusion-Flame Extinction 69
Droplet Burning 74
Flame Spread 78
Current Problems in Diffusion Flames 81
Bibliography 81

4. Flammability, Explosions, and Detonations 83

The Self-Acceleratory Nature of Combustion 83
Initiation of Thermal Explosions 86

Ignition   91
Flammability and Explosion Limits   95
Detonation Structure   98
Detonation Stability   103
Detonation Development and Failure   105
Outstanding Problems in Ignition, Explosions, and Detonations   108
Bibliography   109

5. Turbulent Combustion   111

Regimes of Turbulent Combustion   111
Turbulent Premixed Flames   119
Turbulent Diffusion Flames   132
Approaches to the Theory of Turbulent Combustion   140
Outstanding Problems in Turbulent Combustion   149
Bibliography   150

6. The Future   153

Energy Sources and Explosion Safety   153
Aerospace Propulsion   154
Predictions of Specific Advances in Combustion Science   155
General Considerations in the Science of Combustion   159
Bibliography   161

Index   163

Fundamental Aspects of Combustion

1

BACKGROUND AND
FORMULATION OF COMBUSTION

Although combustion has a long history and great economic and technical importance, its scientific investigation is of relatively recent origin. Combustion science can be defined as the science of exothermic chemical reactions in flows with heat and mass transfer. As such, it involves thermodynamics, chemical kinetics, fluid mechanics, and transport processes. Since the foundations of the second and last of these subjects were not laid until the middle of the nineteenth century, combustion did not emerge as a science until the beginning of the twentieth century. In recent years, great improvements in understanding of combustion processes have arisen through advances in computer capabilities, in experimental techniques, and in asymptotic methods of applied mathematics.

The present volume is intended to summarize the basis of the subject and to indicate areas of current activity—fundamental and applied—as well as to address prospects for future progress. It is hoped that some of the multidisciplinary fascination of combustion can be conveyed to the reader.

History of the Technology and Science of Combustion

Technological developments in an area often precede the emergence of the area as a firmly established science. This seems to have been especially true in combustion, and in many respects it remains true today. Table 1.1 is an approximate chronological list of some of the technological and scientific developments related to combustion. Mythological aspects, such as the stealing of fire from the gods by Prometheus (Fig. 1.1), are excluded here.

Perhaps because fire is so mystifying to the senses, combustion enjoyed more than its fair share of scientific speculation prior to the

Table 1.1. Chronological history of combustion

	Technology	Science
Prehistoric	Fire for warmth and for preparation of food	
100	Metallurgy	Fire is one of four elements
1000–1200	Pyrotechnic rockets, furnaces	
1300	Guns	
1600	Steam engine	*Chemistry*
		Jean Ray, Boyle,
1700	Industrial revolution	Hooke, Mayow (Stahl)
		Lavosier, Priestley, Scheele, Davy
1866		Bunsen
1876	Otto engine	*Deflagration, Detonation*
1881		Berthelot & Vieille
		Mallard & Le Chatelier
1890	Diesel engine	Mikhel'son
		Chapman
1905		Jouguet
		Combustion Science
1913		Taffanel
1928		Burke & Schumann
1930		Daniell
1938, 1940		Zel'dovich & Frank-Kamenetskii
1950, 1960		Computers, asymptotics, lasers

Renaissance. During the seventeenth century, combustion processes were of central importance to the development of the modern science of chemistry, although they also formed the core of Stahl's erroneous phlogiston theory of matter. The experiments of Bunsen and others on the combustion processes of deflagration and detonation led to the emergence of the science around the turn of the twentieth century.

The names of some of the people involved in laying the foundations of the science of combustion are given in Table 1.1. Especially notable contributions came from France, Russia, England, Germany and America. Mathematical definitions of deflagrations and detonations are due to Chapman and Jouguet, while the clearest early explanation of the propagation mechanism of deflagrations was offered by Mikhel'son. Russian leadership in the science continued in the Thirties and Forties

Fig. 1.1. The punishment of Prometheus for giving fire to mankind (by T. Velasquez).

through research in which Zel'dovich was involved; virtually all aspects of combustion theory bear his mark. Today activity extends practically throughout the world, with further important scientific advances having come from the Orient (especially Japan), from the southern hemisphere (notably Australia), and from most countries in Eastern and Western Europe, including Spain.

Applications

Some of the areas of application of combustion science are listed in Table 1.2. In the earliest applications, combustion was used to heat air in furnaces or to heat water in boilers. The heat may be used directly (for warmth, cooking, metal forming, etc.) or may be employed to produce work in engines. In piston engines the work may be done by a fluid different from that in which the combustion occurs (external combustion engines, such as the steam engine), or by the same fluid (internal combustion engines, such as spark-ignition and compression-ignition or Diesel engines). Gas turbines for stationary power plants and for jet aircraft experienced extensive development during the middle part of the present century. Rocket engines, although of ancient origin, were

Table 1.2. Some applications of combustion

1. Furnaces and boilers
2. Piston engines
3. Gas turbines and jet engines
4. Rocket engines
5. Guns and explosives
6. Forming of materials
7. Chemical processing
8. Fire hazards and safety
9. Chemical lasers

advanced at a remarkable pace during the Forties and Fifties, resulting in feasibility of the dream of space travel for mankind (Fig. 1.2).

Explosives are needed in mining and construction. Beyond the refinement of materials by the heat of combustion, for example by use of blast furnaces in the steel industry, there are solids that can burn to solid products, producing potentially useful new materials directly through combustion. Numerous rare but useful chemicals are efficiently produced in controlled combustion processes; even the recently discovered ball-shaped carbon-sixty molecules (fullerenes) have been obtained from the soot produced in flames under suitable conditions.

Chemical lasers can take advantage of nonequilibrium in combustion processes to generate intense beams having various applications. The list of applications of combustion is continually increasing and shows no signs of exhaustion. These applications stem from improved understanding of the basic science, which rests today on a firm set of conservation equations.

Fig. 1.2. A night launch of the space shuttle.

Conservation Equations

In the systematic advancement of a physical science, conservation equations are sought which express, in a mathematically precise manner, relationships among various quantities that occur in the science. In much of physics the conservation equations are not known, and the activities in progress represent attempts to find them. In a mature or applied physical science, the conservation equations are known, and efforts are directed toward solving them for quantities of interest, typically under conditions of practical importance. In this sense, combustion science is a mature science. Its conservation equations are partial differential equations expressing conservation of mass, momentum, energy, and chemical species.

The conservation equations for combustion of gases are those of fluid mechanics, supplemented by equations for conservation of chemical species that include effects of chemical kinetics. For mixtures of ideal gases these are the Navier-Stokes equations augmented by chemical kinetics, which may be derived from the kinetic theory of gases by assuming that molecular mean free paths are small compared with the characteristic length scales of the system and that departures from thermodynamic equilibrium (but not chemical equilibrium) also are small locally. The derivation, through the Chapman-Enskog expansion for small ratios of collision durations to free times, also provides equations for molecular transport and kinetic-theory expressions for transport coefficients. For these reasons, the derivation from kinetic theory is more satisfying than a phenomenological derivation from continuum concepts. On the other hand, for combustion of solids or liquids, the phenomenological derivation is more satisfying because the intermolecular interactions occur continuously and are more complex; in all cases the conservation equations are the same, but the supplementary equations of state and transport equations differ.

Assumptions are that times for equilibration of translational and internal degrees of freedom of molecules are short compared with characteristic times of the processes investigated and with the times for chemical transformations. It is then possible to define a temperature, for example, and to write expressions for rates of chemical changes as functions of temperature and concentrations of chemical species. This level of description has been found most appropriate for combustion processes. In examples in which equilibrium of internal molecular states is not maintained, as in chemical lasers, it is straightforward to employ exactly the same conservation laws but to treat molecules in different internal states as different chemical species. Thus translational equilibrium in elastic collisions is really the only underlying requirement. This requirement is violated in strong shock waves such as those occurring in detonations, and kinetic theory (for example, the Boltzmann equation) is needed for describing the shock structure, but from the viewpoint of combustion the shocks are chemically frozen discontinuities because the chemical reactions require many molecular collisions. An exception may arise in detonation of certain solids, where chemical transformations might occur as fast as translational energy changes; such processes are poorly understood.

One way to exhibit the physical phenomena that arise in combustion is to write the conservation equations in a form in which the variables have been made nondimensional through division by constant characteristic values. Constant dimensional quantities are listed in Table 1.3, and nondimensional variables formed from them are shown in Table 1.4. From the constant dimensional quantities, constant nondimensional parameters can be constructed as shown in Table 1.5; the values of these parameters largely determine the kinds of combustion processes that occur, and many of them are named after scientists who were instrumental in elucidating their meaning.

Table 1.3. Constant dimensional quantities

τ	Characteristic evolution time
ℓ	Characteristic length
v_o	Characteristic velocity
ρ_o	Characteristic density
p_o	Characteristic pressure
T_o	Characteristic temperature
μ_o	Characteristic viscosity
λ_o	Characteristic thermal conductivity
c_{po}	Characteristic specific heat at constant pressure
D_{ijo}	Characteristic binary diffusion coefficient for species pair i, j
D_{Tio}	Characteristic thermal diffusion coefficient for species i in the mixture
q_o	Characteristic radiant energy flux (e.g., $\epsilon_o \sigma T_o^4$, ϵ = emissivity, σ = Stefan-Boltzmann constant)
g_o	Characteristic body force per unit mass
h_o	Characteristic heat release per unit mass of mixture
E_k	Activation energy for forward reaction k
H_k	Heat release in forward reaction k
R_o	Universal gas constant
τ_{fk}	Characteristic time for forward reaction k
τ_{bk}	Characteristic time for backward reaction k

Table 1.4. Nondimensional variables

t	Time over τ
$\boldsymbol{x}$	Space coordinates over ℓ
$\boldsymbol{v}$	Velocity vector over v_o
ρ	Density over ρ_o
p	Pressure over p_o
T	Temperature over T_o
X_i	Mole fraction of species i
Y_i	Mass fraction of species i ($W_i X_i / \sum_{j=1}^{N} W_j X_j$, W_i = molecular weight of species i)
$\boldsymbol{V}_i$	Diffusion velocity of species i over v_o
$\boldsymbol{f}_i$	Body force per unit mass on species i over g_o $\left(\boldsymbol{f} = \sum_{i=1}^{N} Y_i \boldsymbol{f}_i \right)$
T	Shear-stress tensor over $(\mu_o v_o / \ell_o)$
$\boldsymbol{q}$	Radiant energy flux over q_o
λ	Thermal conductivity over λ_o
μ	Coefficient of viscosity over μ_o
$d_{ij}(p, T)$	Binary diffusion coefficient of species pair i, j over D_{ijo}
$e_{ij}(p, T, X_1, \dots, X_N)$	Thermal diffusion coefficient of species i over D_{Tio}, divided by $\rho d_{ij}(p, T)$
$f_k(p, T)$	Nondimensional pressure and secondary temperature dependences of forward rate of step k
$g_k(p, T)$	Nondimensional pressure and secondary temperature dependences of equilibrium constant for step k
c_p	Specific heat of mixture over c_{po}
h_i	Enthalpy of formation per unit mass for species i at temperature T_o over h_o

The nondimensional conservation equations can be written as:

Mass Conservation

$$\omega \frac{\partial \rho}{\partial t} + \boldsymbol{\nabla} \cdot (\rho \boldsymbol{v}) = 0 \tag{1.1}$$

Momentum Conservation

$$\omega \frac{\partial (\rho \boldsymbol{v})}{\partial t} + \boldsymbol{\nabla} \cdot (\rho \boldsymbol{v} \boldsymbol{v}) = -\frac{\boldsymbol{\nabla} p}{M^2} + \frac{\rho \boldsymbol{f}}{F} + \frac{\boldsymbol{\nabla} \cdot \mathsf{T}}{R} \tag{1.2}$$

Energy Conservation

$$\left(\omega \rho \frac{\partial}{\partial t} + \rho \boldsymbol{v} \cdot \boldsymbol{\nabla} \right) \left(\int c_p dT + \alpha \sum_{i=1}^{N} h_i Y_i + \frac{\gamma - 1}{\gamma} M^2 \frac{v^2}{2} \right)$$

$$= \frac{\gamma - 1}{\gamma} \omega \frac{\partial p}{\partial t} + \frac{\boldsymbol{\nabla} \cdot (\lambda \boldsymbol{\nabla} T)}{PR} - \alpha \sum_{i=1}^{N} h_i \left[\boldsymbol{\nabla} \cdot (\rho Y_i \boldsymbol{V}_i) \right]$$

Table 1.5. Nondimensional parameters

α	$h_o/(c_{po}T_o)$ Nondimensional heat release
β_k	$E_k/(R_oT_o)$ Nondimensional activation energy for reaction k
γ_k	$H_k/(R_oT_o)$ Nondimensional heat release for reaction k
D_k	$\ell/(v_o\tau_{fk})$ Damköhler's first similarity group for reaction k
K_k	τ_{bk}/τ_{fk} Nondimensional equilibrium constant for reaction k
ν_{ik}	Increase in number of moles of species i in reaction k (stoichiometric coefficient)
n_{ik}	Order of forward reaction k with respect to species i
m_{ik}	Order of backward reaction k with respect to species i
P	$\mu_o c_{po}/\lambda_o$ Prandtl number
S_{ij}	$\mu_o/(\rho_o D_{ijo})$ Schmidt number for species pair i,j
α_{ij}	$D_{Tio}/(\rho_o D_{ijo})$ Thermal-diffusion ratio for species i and species pair i,j
B	$\rho_o v_o c_{po} T_o/q_o$ Boltzmann number
γ	Ratio of specific heat at constant pressure to specific heat at constant volume (often a parameter but could be a variable)
M	$v_o/\sqrt{p_o/\rho_o}$ Newtonian Mach number
F	$v_o^2/(g_o\ell)$ Froude number
R	$\rho_o v_o\ell/\mu_o$ Reynolds number
ω	$\ell/(v_o\tau)$ Ratio of flow time to evolution time

$$+\frac{\nabla\cdot\boldsymbol{q}}{B}+\frac{\gamma-1}{\gamma}\frac{M^2}{R}\nabla\cdot(\boldsymbol{v}\cdot\mathsf{T})+\frac{\gamma-1}{\gamma}\frac{M^2}{F}\left(\rho\boldsymbol{v}\cdot\boldsymbol{f}-\rho\sum_{i=1}^{N}Y_i\boldsymbol{V}_i\cdot\boldsymbol{f}_i\right)\quad(1.3)$$

Species Conservation

$$\omega\frac{\partial(\rho Y_i)}{\partial t}+\nabla\cdot(\rho\boldsymbol{v}Y_i)=-\nabla\cdot(\rho Y_i\boldsymbol{V}_i)+\sum_{k=1}^{L}\nu_{ik}D_k f_k(p,T)e^{-\beta_k/T}$$

$$\times\prod_{j=1}^{N}X_j^{n_{jk}}\left[1-\frac{g_k(p,T)}{K_k}e^{-\gamma_k/T}\prod_{j=1}^{N}X_j^{m_{jk}-n_{jk}}\right],\quad i=1,\ldots,N\quad(1.4)$$

where there are N different chemical species, and L different chemical reactions may occur. In the problems encountered in theoretical analyses of combustion processes, these conservation equations, subject to suitable initial and boundary conditions, are to be solved for $\rho(\boldsymbol{x},t)$, $\boldsymbol{v}(\boldsymbol{x},t)$, $T(\boldsymbol{x},t)$, and $Y_i(\boldsymbol{x},t)$.

The differential equation for mass conservation (1.1) expresses a balance between the rate of accumulation of mass within a volume element and the rate at which mass flows out of that element; since the fluid velocities are very small compared with the velocity of light, mass and energy are conserved separately. Chemical reactions, which rearrange atoms among molecules, neither create nor destroy mass.

Therefore (1.1) continues to apply in the presence of chemistry, and the summation of (1.4) over all species i, from $i = 1$ to $i = N$, gives (1.1). This last result readily follows mathematically from the definitions of diffusion velocities as the velocities with respect to the mass-average velocity of the fluid ($\sum_{i=1}^{N} \rho Y_i V_i = 0$) and from mass conservation in chemical conversion at the microscopic level ($\sum_{i=1}^{N} \sum_{k=1}^{L} \cdots = 0$). It implies that (1.1) and (1.4) together constitute only N (not $N + 1$) independent equations, reflecting the identity $\sum_{i=1}^{N} Y_i = 1$, by definition.

The four different types of terms appearing in (1.4) lie at the heart of the science of combustion and describe accumulation of a chemical species within a volume element, convection (also called advection) of this chemical species out of the volume element by the mass-average motion of the fluid, diffusion of the species into the volume element, and production of the species within the element by chemical reactions. This accumulation-convection-diffusion-production balance is the most general conservation statement for fluid mixtures composed of molecules subject to chemical conversion.

The terms in the equation for momentum conservation (1.2) express a balance of the rate of increase of momentum, inertial forces, pressure forces, body forces, and viscous forces, respectively. Magnitudes of the last three of these are measured by the Mach number M, the Froude number F and the Reynolds number R.

In the energy conservation equation (1.3), three kinds of energy are seen to arise—thermal energy, $\int c_p dT$; chemical energy, $\alpha \sum_{i=1}^{N} h_i Y_i$; and kinetic energy, $(\gamma - 1/\gamma)M^2(v^2/2)$. The second of these, which is special to combustion and is not usually addressed in fluid mechanics, is associated with the different binding energies of different molecules in their ground states. Through (1.4) it provides a chemical-kinetic contribution to the change in thermal energy (more precisely, enthalpy) of the material. The same four types of processes that appear in the balance (1.4) therefore also occur in (1.3), where the product of the Prandtl and Reynolds numbers PR, also called a Peclet number, measures the magnitude of the diffusive effects (here heat conduction). There can be additional transient energy accumulation through pressure increase, $(\gamma - 1/\gamma)\omega(\partial p/\partial t)$, of importance in internal combustion engines, as well as radiant energy input (measured by the Boltzmann number B), and surface-force and body-force work (in the last line) that affect energy conservation. Therefore, in general, more phenomena must be addressed in describing energy conservation than species conservation.

The conservation equations can be solved only if necessary auxiliary relationships are available for quantities appearing on the right-hand side. Thus, in (1.2), expressions are needed for pressure, body force, and viscous stress. These may be provided by the equation of state for an ideal gas mixture,

$$p = \rho R_o T / \sum_{i=1}^{N} X_i W_i , \qquad (1.5)$$

by formulas for $\boldsymbol{f}_i$ describing gravitational or electrostatic effects, and by an expression for the viscous stress tensor. For a Newtonian fluid,

$$\mathsf{T} = \mu \left[(\boldsymbol{\nabla v}) + (\boldsymbol{\nabla v})^T \right] - \mu' (\boldsymbol{\nabla \cdot v}) \mathsf{U}, \qquad (1.6)$$

where the superscript T denotes the transpose of the tensor; U is the unit tensor; and μ' a nondimensional second viscosity coefficient, $\frac{2}{3}\mu$ minus a nondimensional coefficient of bulk viscosity (the latter zero for monatomic gases). For chemically reacting ideal gas mixtures the kinetic theory of gases provides μ, μ', and λ as functions of T, X_1, ..., X_N. In energy conservation (1.3), caloric equations of state give c_p; for ideal gas mixtures, $c_p = \sum_{i=1}^{N} Y_i c_{p_i}$ with $c_{p_i}(T)$ obtained from the structure of the molecule.

In general, an equation for transport of radiation must be added to determine $\boldsymbol{q}$ in (1.3). This brings in the field of radiation gasdynamics and converts the system of differential equations to one of integrodifferential equations, since the scattering term in the transport-emission-absorption-scattering balance of radiation transport involves an integral. Although radiation often is important in combustion, for example in providing the main heat flux to the walls of large furnaces, frequently absorption and scattering can be neglected, so that $\boldsymbol{\nabla \cdot q} = \epsilon T^4$ (with $\epsilon \equiv \ell_{P_o}/\ell_P$, $\epsilon_o \equiv 4\ell/\ell_{P_o}$, in terms of a Plank-mean absorption length ℓ_P), which removes the necessity for analysis of radiation transport.

In (1.3) and (1.4) the diffusion velocities V_i appear. These are obtained from a diffusion equation, the general form of which is found from the kinetic theory of gases to be

$$R \sum_{j=1}^{N} \frac{S_{ij}}{d_{ij}(p,T)} X_i X_j (\boldsymbol{V}_j - \boldsymbol{V}_i) = \boldsymbol{\nabla} X_i + (X_i - Y_i) \frac{\boldsymbol{\nabla} p}{p}$$

$$+ \frac{M^2}{F} \frac{\rho}{p} \sum_{j=1}^{N} Y_i Y_j (\boldsymbol{f}_j - \boldsymbol{f}_i) + \sum_{j=1}^{N} X_i X_j \left[\frac{\alpha_{ij} e_{ij}(p,T,X_1,\ldots,X_N)}{Y_i} \right.$$

$$\left. - \frac{\alpha_{ji} e_{ji}(p,T,X_1,\ldots,X_N)}{Y_j} \right] \frac{\boldsymbol{\nabla} T}{T}, \qquad i = 1,\ldots,N. \qquad (1.7)$$

Equation (1.7) expresses differences in diffusion velocities in terms of concentration gradients, pressure gradients ("pressure-gradient diffusion"), body forces ("body-force diffusion") and temperature gradients ("thermal diffusion"). Often the last three of these effects are negligible so that on the right of (1.7) only the first term survives, giving a set often called the Stefan-Maxwell equations; if also $S_{ij}/d_{ij}(p, T) = S/d(p, T)$, independent of the species pair (i, j), then Fick's law

$$RSY_iV_i = -d(p, T) \nabla Y_i, \qquad i = 1, \ldots, N, \qquad (1.8)$$

is obtained. This simplification, expressing the diffusion velocity as proportional to the concentration gradient of the species diffusing (in the same way that Fourier's law expresses the heat flux as proportional to the temperature gradient), applies for example if one of the chemical species is present in great excess, but in general will be inaccurate in multicomponent flows. The kinetic theory of multicomponent gases provides formulas needed for d_{ij} and e_{ij}.

With the information summarized here, there are a sufficient number of auxiliary relationships to produce a closed system of differential equations for describing combustion phenomena. The complexity of the system of equations exceeds that encountered in many other fields. Moreover, if aspects of two-phase flow and of turbulence are taken into account (see Chapters 3 and 5, respectively), the complexity increases much further. The apparent complexity implies a richness that has attracted many investigators to the field and that has motivated efforts in computational methods for computer implementation. However, in science, understanding is achieved through simplification. There are practically no combustion problems in which all of the effects exhibited here simultaneously play roles. By neglecting unimportant phenomena, for example through exercise of asymptotic methods, sufficiently accurate descriptions of combustion processes can be obtained with transparent interpretations. An objective here is to expose some of these simpler physical interpretations of combustion phenomena. For this purpose, classifications of different types of combustion processes are helpful.

Classifications of Combustion Processes

There are a number of ways to place combustion processes into different categories. One approach is to investigate limits of values of parameters in the conservation equations. Some limits are listed in Table 1.6. Many problems involve steady flows, $\omega = 0$, as established, for example, in wind tunnels or in continuous-feed reactors (as opposed to batch reactors). The limit $\omega \to 0$ may be called that of slowly evolving flames. Although there is renewed interest in supersonic combustion for applications in hypersonic propulsion (aerospace plane), the majority of applications involve low Mach numbers ($M \to 0$), a limit which eliminates acoustics by replacing p by a function of time plus M^2 times a function of x and t in (1.2) and the other equations. By contrast, flows with low, moderate, and high Reynolds numbers all are of interest in different contexts. At sufficiently small scales buoyancy forces usually are negligible ($F \to \infty$). Often the radiant energy flux and thermal diffusion are negligible, the former frequently even when the light output from the combustion process is significant; the light may be chemiluminescence, nonequilibrium radiation giving flames characteristic colors but involving only a small fraction of the total energy present.

In some of the limits it is necessary merely to exclude the corresponding terms from the conservation equations. In others, these terms with small coefficients still may be important in some regions (as in boundary-layer theory for $R \to \infty$) but are negligible in others. This situation calls for the introduction of asymptotic methods in applied mathematics. The parameters for which asymptotics are needed quite often differ in combustion from other sciences. The parameters shown at the end of Table 1.6 are some of the special parameters of combustion.

When the Lewis numbers $L_{ij} \equiv S_{ij}/P$, the ratio of the Schmidt numbers to the Prandtl number, take on the value unity, there is a similarity between energy and species transport that effects a significant simplification in the description of combustion processes. Small departures from this limit, termed nearly equidiffusional flames, can introduce new effects, for example in connection with stability, that are not present when $L_{ij} = 1$. Therefore there is interest in asymptotic analyses for $L_{ij} \to 1$.

Combustion reactions often go nearly to completion, but not quite ($K_k \to \infty$), and perturbations about this limit are of interest. The limit of fast chemistry, described by allowing Damköhler numbers to

Table 1.6. Types of limits

ω	$\to$ 0	Quasisteady flow
M	$\to$ 0	Low-speed flow
F	$\to \infty$	Negligible buoyancy
R	$\to$ 0; ∞	Lubrication theory; Boundary-layer theory
B	$\to \infty$	Negligible radiant energy flux
α_{ij}	$\to$ 0	Negligible thermal diffusion
$L_{ij} \equiv S_{ij}/P$	$\to$ 1	Nearly equal molecular diffusivities
K_k	$\to \infty$	Nearly irreversible reactions
D_k	$\to \infty$	Fast chemistry
γ_k	$\to \infty$	Large heat release
β_k	$\to \infty$	Large activation energy (leads to AEA $\equiv$ activation-energy asymptotics)

become large ($D_k \to \infty$), usually is a singular limit in combustion that needs asymptotic analysis, called Damköhler-number asymptotics (DNA). Sometimes ratios of Damköhler numbers for different chemical steps k are large or small, and asymptotic analyses based either on the presumption that such a ratio approaches zero or infinity, or on the supposition that various key reaction rates are large, as in DNA, are termed rate-ratio asymptotics (RRA). Perhaps the two most important parameters in combustion are the nondimensional heat release γ_k and the nondimensional activation energy β_k (characterizing the temperature dependence of the rate of heat release). Both often are large in combustion. In recent years consideration of the limit $\beta_k \to \infty$ has played a central role in combustion theory, greatly improving our understanding of combustion processes. This limit is singular and is called activation-energy asymptotics (AEA). The product $\gamma_k \beta_k$ has recently been named the Zel'dovich number in honor of Zel'dovich's many contributions to the theory; $\gamma_k \beta_k$ is large in the majority of combustion processes.

Another approach to the classification of combustion processes is based on identifying different flow configurations. In this approach, the most fundamental distinction in combustion is that between *premixed* and *nonpremixed* systems. Although some fuels, such as homogeneous solid propellants and explosives (for example, trinitrotoluene, TNT), can burn exothermically by themselves, most require an oxidizer, as does natural gas (methane);

$$CH_4 + 2O_2 \to CO_2 + H_2O \tag{1.9}$$

is the overall reaction for complete methane oxidation. The character of the combustion process depends strongly on whether the fuel and oxidizer are mixed before initiation of the chemical reactions and fluid flow. An example of premixed combustion is the flame cone of a Bunsen burner, in which the gaseous fuel is thoroughly mixed with air at the base of the burner tube that holds the flame at its upper exit. This is shown as the inner cone in the photograph on the left in Fig. 1.3; residual fuel burns in the outer cone. If the hole that admits air at the base of the tube is closed, then the combustion becomes nonpremixed and occurs as the fuel leaving the tube mixes with air (Fig. 1.3 on the right); the flame is much longer and more yellow with less blue.

Premixed combustion can proceed either in transient processes nearly homogeneously throughout the entire system or in waves (deflagrations or detonations)—thin fronts that propagate into the unburnt combustibles. In contrast, nonpremixed combustion is not intrinsically propagatory and occurs in a flame into which fuel and oxidizer are transported from opposite sides. Since diffusion is essential to this type

Fig. 1.3. A methane flame on a Bunsen burner; premixed (on the left) and nonpremixed (on the right) (photographs by S.-C. Li).

of combustion, nonpremixed flames often are called diffusion flames. They tend to be nonexplosive because heat-release rates are limited by diffusion rates.

In the classification employed in the following chapters, premixed flames (deflagrations) are considered first (Chapter 2) and diffusion flames next (Chapter 3). Ignition and explosion processes that occur in premixed systems with exothermic chemistry, as well as detonations, are then addressed (Chapter 4). Complications associated with turbulence are largely excluded from these discussions. The intrigues of both turbulent premixed flames and turbulent diffusion flames are approached in Chapter 5.

The early ideas of Chapman and Jouguet concerning premixed combustion waves are essential to classifications in distinguishing deflagrations from detonations. Therefore, in the final section of the present chapter, the overall description of steady, planar, combustion fronts is considered.

Deflagrations and Detonations

For steady, one-dimensional flow of a combustible gas that burns to completion, equations relating initial and final conditions are readily derived from (1.1) through (1.3) by use of the requirement of uniformity $(d/dx) = 0$ in the upstream ($x = -\infty$, subscript 0) and downstream

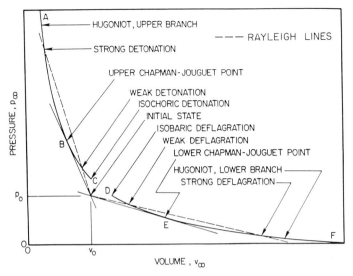

Fig. 1.4. Schematic locus of burnt-gas states for combustion waves.

($x = +\infty$, subscript ∞) mixtures. These equations, called the Rankine-Hugoniot relations, provide jump conditions across the front. With v being the x component of $\boldsymbol{v}$ (the other components vanishing or, more generally, keeping constant values), integration of (1.1) gives

$$m_o \equiv \rho_o v_o = \rho_\infty v_\infty. \tag{1.10}$$

Similarly, integration of (1.2) with $\boldsymbol{f} = 0$ and T vanishing upstream and downstream yields

$$P_o \equiv \rho_o v_o^2 + p_o = \rho_\infty v_\infty^2 + p_\infty. \tag{1.11}$$

Contrary to the formulation given previously, all symbols represent dimensional quantities throughout the present section.

The sequence of final states obeying

$$p_\infty + m_o^2/\rho_\infty = P_o \equiv p_o + m_o^2/\rho_o, \tag{1.12}$$

obtained by substituting (1.10) into (1.11), is the Rayleigh line, a straight line in the plane of final pressure, p_∞, and specific volume, $1/\rho_\infty$, such as Fig. 1.4.

Integration of (1.3) gives, in dimensional variables,

$$\int_{T_0}^{T_\infty} c_p dT - h_o + \frac{1}{2} m_o^2 \left(\frac{1}{\rho_\infty^2} - \frac{1}{\rho_o^2} \right) = 0, \tag{1.13}$$

where use has been made of (1.10). Here h_o is the total amount of chemical heat release per unit mass of the mixture, and the final temperature T_∞ is the adiabatic flame temperature if the terms involving m_o^2 are negligible so that pressure is nearly constant. Use of (1.12) in (1.13) to eliminate m_o^2 provides a relationship among thermodynamic properties, the Hugoniot curve, which can be written as

$$\int_{T_0}^{T_\infty} c_p dT - h_o = \frac{1}{2}\left(\frac{1}{\rho_\infty} + \frac{1}{\rho_o}\right)(p_\infty - p_o). \tag{1.14}$$

If the ratio γ of specific heats is constant, then with the aid of the equation of state (1.5), this formula can be written explicitly as

$$\left(\frac{\gamma}{\gamma - 1}\right)\left(\frac{p_\infty}{\rho_\infty} - \frac{p_o}{\rho_o}\right) - \frac{1}{2}\left(\frac{1}{\rho_\infty} + \frac{1}{\rho_o}\right)(p_\infty - p_o) = h_o. \tag{1.15}$$

The Hugoniot curve is shown schematically in Fig. 1.4 for a representative combustion system. The final state is determined by the intersection of the Rayleigh line with the Hugoniot curve, as illustrated.

For $h_o > 0$ it is seen that in the first quadrant there is no intersection of the Rayleigh line with the Hugoniot curve for real values of m_o. This excluded quadrant divides the Hugoniot into two branches, an upper branch of large ρ_∞ and p_∞, called the detonation branch, and a lower branch of small ρ_∞ and p_∞, called the deflagration branch. Deflagrations are combustion waves with end states on the lower branch. Their structures will be seen to differ in numerous ways from those of detonations, the combustion waves with end states on the upper branch. As h_o increases, the separation between the initial state and the Hugoniot increases.

Since the negative of the slope of the straight line connecting the burnt state to the initial state in Fig. 1.4 is proportional to the square of the propagation velocity of the wave, it is seen that detonations propagate faster than deflagrations. There is a minimum propagation velocity for detonations, corresponding to tangency at the upper Chapman-Jouguet point, and a maximum propagation velocity for deflagrations, corresponding to tangency at the lower Chapman-Jouguet point. The two dashed lines in Fig. 1.4 illustrate representative intermediate conditions, and each has two intersections, corresponding to weak and strong waves, as indicated in the figure.

The waves encountered in combustion are weak (in fact, nearly isobaric) deflagrations and strong or often Chapman-Jouguet detonations, as discussed in later chapters. Some additional properties of these waves

Table 1.7. Properties of Hugoniot curves

Name	Section in Fig 1.4	Pressure ratio $p = (p_\infty/p_o)$	Volume ratio $v = (v_\infty/v_o)$	Propagation Mach number M_o	Burnt-gas Mach number M_∞	Remarks
Strong detonations	Line $A - B$	$p_+ < p < \infty$	$v_{\min} < v < v_+$ $(v_{\min} > 0)$	$M_{o+} < M_o < \infty$	$M_\infty < 1$	Seldom observed; requires special experimental arrangement.
Upper Chapman-Jouguet point	Point B	$p = p_+$ $(p_+ > 1)$	$v = v_+$ $(v_+ < 1)$	$M_o = M_{o+}$ $(M_{o+} > 1)$	$M_\infty = 1$	Usually observed for waves propagating in tubes.
Weak detonations	Line $B - C$	$p_1 < p < p_+$ $(p_1 > 1)$	$v_+ < v < 1$	$M_{o+} < M_o < \infty$	$M_\infty > 1$	Seldom observed; requires very special gas mixtures.
Weak deflagrations	Line $D - E$	$p_- < p < 1$	$v_1 < v < v_-$ $(v_1 > 1)$	$0 < M_o < M_{o-}$	$M_\infty < 1$	Often observed; $p \approx 1$ in most experiments.
Lower Chapman-Jouguet point	Point E	$p = p_-$ $(p_- < 1)$	$v = v_-$ $(v_- > 1)$	$M_o = M_{o-}$ $(M_{o-} < 1)$	$M_\infty = 1$	Not observed.
Strong deflagrations	Line $E - F$	$0 < p < p_-$	$v_- < v < v_{\max}$ $(v_{\max} < \infty)$	$M_{o\min} < M_o < M_{o-}$ $(M_{o\min} > 0)$	$M_\infty > 1$	Not observed; forbidden by considerations of wave structure.

are summarized in Table 1.7, where the subscripts $+$ and $-$ refer to upper and lower Chapman-Jouguet conditions, respectively, and the subscripts 1, max, and min identify additional constants determined by the initial state of the mixture.

Bibliography

Buckmaster, J.D. and Ludford, G.S.S. 1982. *Theory of laminar flames.* Cambridge University Press, Cambridge, England.

Glassman, I. 1987. *Combustion,* 2nd ed. Academic Press, New York.

Strehlow, R.A. 1984. *Combustion fundamentals.* McGraw-Hill, New York.

Williams, F.A. 1985. *Combustion theory,* 2nd ed. Addison-Wesley Publishing Company, Menlo Park, California.

Williams, F.A. 1987. Combustion. *Encyclopedia of physical science and technology,* vol. 3. Academic Press, New York.

Zel'dovich, Y.B., Barenblatt, G.I., Librovich, V.B., and Makhviladze, G.M. 1980. *The mathematical theory of combustion and explosion.* Nauka, Moscow; English translation, 1985, Consultants Bureau, A Division of Plenum Publishing Corporation, New York.

2

PREMIXED FLAMES

The deflagrations defined in the last section of Chapter 1 also are called premixed flames and occur in many combustion experiments. If a uniform combustible mixture contained in a tube is ignited by a spark, then a premixed flame propagates through the mixture, moving along the tube at a fairly definite velocity characteristic of the mixture. Such deflagration waves can be made stationary in the laboratory by causing the combustible mixture to flow at an appropriate velocity. This is achieved in a Bunsen burner, where the deflagration is anchored at the exit of the tube, which then serves as a flame holder.

Deflagrations propagate at a velocity that depends on the pressure, the initial temperature of the combustible mixture, and its chemical composition. This velocity is called the deflagration velocity or the laminar burning velocity. In a Bunsen burner it can be measured as the sine of the half-angle of the flame cone multiplied by the exit velocity of the gas leaving the tube. Thus, increasing the flow velocity in the tube causes the cone to elongate and its angle to decrease. If the flow velocity is decreased below the burning velocity, the flame first becomes flat and then propagates down the tube (flashback). Because of flow nonuniformities, the flat flame is not stable on the Bunsen burner. However, there are ways to stabilize flat premixed flames. One is to pass the combustible through a cooled, porous, flat plate (the flat-flame burner) and ignite it on the downstream side; because of heat loss to the burner, the flat flame is stable over a range of flow velocities, and it propagates downstream (blowoff) only when the flow velocity exceeds the burning velocity. Another approach (the stagnation-flow burner) is to establish counterflowing streams of the combustible, or of the combustible and its reaction products, so that the flat flame can exist when the exit velocity of the stream exceeds the burning velocity.

Experiments of these types have been employed to measure burning velocities of many combustible mixtures. Representative results for a

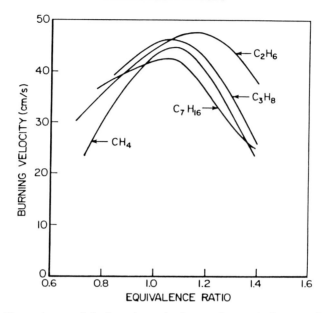

Fig. 2.1. Dependence of the burning velocity on the equivalence ratio for four hydrocarbon fuels in air at standard conditions.

number of hydrocarbon-air mixtures at atmospheric pressure and room temperature are shown in Fig. 2.1, where the equivalence ratio is defined as the fuel-air ratio of the mixture divided by the stoichiometric fuel-air ratio (the ratio needed for complete combustion to CO_2 and H_2O). This chapter concerns the prediction of curves like those in Fig. 2.1 and the internal structures of the deflagrations that exhibit these burning velocities.

Simplified Description of Flame Structure

The clearest early description of deflagration structure was offered by Mikhel'son (1889). The pressure p is approximated as constant, and with the specific heat c_p of the mixture taken as a constant, the overall energy conservation shows that the energy per unit mass added to the mixture by combustion is

$$h_o = c_p \left(T_\infty - T_o \right), \tag{2.1}$$

where T_o is the initial temperature and T_∞ the final temperature (see 1.14). This energy is added at a finite rate determined by the rate of conversion of reactant mass to product mass. In a one-step, Arrhenius approximation, this conversion rate (mass per unit volume per unit time) can be written as

$$w = AY_F^m Y_O^n e^{-E/(R_o T)}, \qquad (2.2)$$

where the subscripts F and O identify fuel and oxidizer, the mass fractions Y_i are proportional to the concentrations (the number of molecules per unit volume) of the reactants, and m and n are overall reaction orders with respect to fuel and oxidizer, respectively. It is qualitatively correct to assume that these orders, the overall activation energy E, and the prefactor A, are constants characteristic of the combustible mixture. In terms of the thickness δ of the flame, the chemical energy released per unit area per unit time may be approximated as

$$h_o w \delta = \lambda (T_\infty - T_o) / \delta, \qquad (2.3)$$

where λ is the thermal conductivity. Equation (2.3) says that the heat release produces upstream conduction of energy from the hot products to the cool reactants. The conduction is needed to heat the gas to a temperature at which the rate w is significant. The simultaneous importance of heat conduction and finite-rate chemistry in energy conservation was known to Mikhel'son and is central to deflagration structure.

Use of (2.1) to solve (2.3) for δ gives

$$\delta = \sqrt{\lambda/(c_p w)}, \qquad (2.4)$$

showing that the flame thickness varies inversely as the square root of the reaction rate. The mass of reactant converted per unit area per unit time is $\rho_o v_o$, where ρ_o is the initial density of the mixture and v_o the laminar burning velocity. By conservation of fuel mass in the steady flow, the estimate

$$\rho_o v_o = w \delta \qquad (2.5)$$

is obtained. Use of (2.4) in (2.5) gives the burning-velocity formula of Mikhel'son,

$$v_o = \sqrt{w \lambda/c_p} / \rho_o, \qquad (2.6)$$

showing that the velocity is proportional to the square root of the ratio of the thermal diffusivity, $\lambda/(c_p \rho_o)$, to the reaction time, ρ_o/w. This correct prediction of functional dependences represents a remarkable success at so early a stage in the development of the science.

It seems logical to extend the reasoning to recognize a balance between production of reaction products and upstream diffusion of reaction products. If D is the coefficient of molecular diffusion between reactants and products, then as in (2.3) this balance gives $w \delta = \rho_o D/\delta$, so that $\delta = \sqrt{\rho D / w}$. If the Lewis number, $L \equiv \lambda / (c_p \rho_o D)$ is unity the result is the same as (2.4), but often Lewis numbers differ somewhat from unity, so that the results differ. When they differ it is tempting to employ an average diffusivity $\bar{D} \equiv [D + \lambda / (c_p \rho_o)] / 2$, in terms of which (2.6) becomes

$$v_o = \sqrt{\bar{D}/\tau_c}, \qquad (2.7)$$

where $\tau_c \equiv \rho_o / w$ is a chemical time. This reasoning is wrong in that for (2.7) to hold, the definition $\bar{D} \equiv [\lambda / (c_p \rho_o)]^2 / D$ must be introduced; molecular diffusion retards rather than promotes the burning velocity. To see why, it is necessary to augment the reasoning of Mikhel'son.

The flame structure according to Mikhel'son's view is illustrated in Fig. 2.2a, where $Y \equiv Y_F / Y_{Fo}$, in which Y_{Fo} is the initial fuel mass fraction. Fuel-lean or stoichiometric conditions are considered, so that no fuel is left in the final mixture ($Y_{F\infty} = 0$). In Fig. 2.2a the entire flame, of thickness δ, is approximated as having a constant rate w. However, according to (2.2), the rate depends strongly on T because typical values of E/R_o, 10,000 K to 30,000 K, are large compared with all values of T within the flame (typically the maximum is $T_\infty \approx 2000$ K). The strong temperature dependence underlies the AEA methods identified in the preceding chapter. The strength of the temperature dependence is measured by the Zel'dovich number

$$\beta = E\,(T_\infty - T_o)\,/\left(R_o T_\infty^2\right), \qquad (2.8)$$

which was denoted by $\gamma_k \beta_k$ for an individual reaction k in the Combustion Processes section of Chapter 1. The main idea is that throughout most of the thickness δ the rate w is negligible; only in a small part of the flame, at the hot end, is w appreciable. The thickness of the reaction zone is then δ / β, as illustrated in Fig. 2.2b, where the asterisk identifies the beginning of the reaction zone. Since typically $\beta \approx 10$, the contribution of the reaction zone to the flame thickness may be neglected in a first approximation, and therefore the thickness δ shown in Fig. 2.2b excludes this contribution for simplicity.

Recall from the discussion of (1.4) that the four effects in the conservation equations are accumulation, convection, diffusion, and reaction. Since the flow under consideration is steady, accumulation is absent. Where the reaction rate is negligible there therefore exists a balance between convection and diffusion. Thus, the part of the flame upstream from the reaction zone maintains a balance between these two effects and may be called a convective-diffusive zone, as indicated in Fig. 2.2b. This zone often is called, more briefly, the preheat zone since it is the

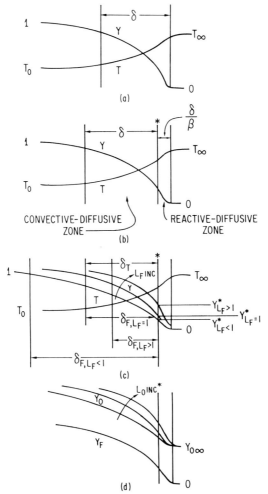

Fig. 2.2. Illustration of the structure of premixed laminar flames (a) according to Mikhel'son, (b) according to AEA, (c) showing the effect of the Lewis number of the fuel, (d) showing the effect of the Lewis number of the oxidizer.

zone in the deflagration in which the reactants are heated to a temperature at which they begin to react at an appreciable rate. The reaction zone is identified in Fig. 2.2b as a reactive-diffusive zone because it is found that in this zone, in a first approximation, there is a balance between the reaction and diffusion terms, the effects of convection there being smaller than these terms.

In the improved model, the rate of energy release per unit area in the reaction zone is equated to that conducted upstream in the preheat zone. The balance in (2.3) thus is revised to be

$$h_o w \left(\delta/\beta\right) = \lambda \left(T_* - T_o\right)/\delta \approx \lambda \left(T_\infty - T_o\right)/\delta , \qquad (2.9)$$

in which the last equality follows from the approximate equality of the upstream temperature T_* and the downstream temperature T_∞ of the reaction zone. In (2.4), the ratio β of the zone thicknesses now appears,

$$\delta = \sqrt{\lambda\beta/(c_p w)}, \qquad (2.10)$$

and the balance (2.5) must now read

$$\rho_o v_o = w \left(\delta/\beta\right), \qquad (2.11)$$

so that (2.6) should become

$$v_o = \sqrt{w\lambda/(c_p\beta)}/\rho_o . \qquad (2.12)$$

Note that because the reaction zone is now smaller, the flame thickness is larger and the burning velocity smaller.

In this structure it is clear from Fig. 2.2b that the fuel concentration in the reaction zone is smaller than its initial concentration because of the convective-diffusive balance for fuel in the preheat zone. Therefore, from (2.2), w is smaller in the reaction zone than the value obtained from an estimate in which $Y_F \approx Y_{Fo}$ is employed. To obtain a reasonable numerical value for w, $T \approx T_o$ cannot be employed in (2.2); this substitution would give the rate in the cold, fresh mixture, which is negligibly small (the fresh mixture being in a metastable state). It is most convenient to employ T_∞, Y_{Fo} and Y_{Oo} in (2.2) to estimate a value for w, identified here as $w_{o\infty}$. Then, in (2.10) and (2.12), with the change in Y_O neglected, $w = w_{o\infty}/\beta^m$, which accounts for the reduction in the reaction rate through the diffusive depletion of fuel.

This reasoning does not yet determine the $\bar{D}$ in (2.7) because it has not addressed influences of differing coefficients of diffusion. The conservation equations (1.3) and (1.4) in the convective-diffusive zone can be written approximately as

$$v\frac{dT}{dx} = D_T \frac{d^2 T}{dx^2} , \qquad v\frac{dY_F}{dx} = D_F \frac{d^2 Y_F}{dx^2} , \qquad (2.13)$$

where D_F is the diffusion coefficient for fuel and $D_T \equiv \lambda/(\rho c_p)$ is the thermal diffusivity. Solutions to these equations are readily found to be approximately of the form

$$T = T_o + (T_\infty - T_o) e^{vx/D_T}, \qquad Y_F = Y_{Fo}\left(1 - e^{vx/D_F}\right), \qquad (2.14)$$

which shows that the ratio of the characteristic length for fuel to that for temperature is $(D_F/v)/(D_T/v) = 1/L_F$, where $L_F \equiv D_T/D_F$ is the Lewis number for the fuel. As illustrated in Fig. 2.2c, this result shows that if the flame thickness δ_T associated with the temperature profile is considered to be held fixed and the Lewis number L_F is allowed to change, then the thickness δ_F associated with the fuel-concentration profile is reduced if $L_F > 1$ and increased if $L_F < 1$; stated simply, the characteristic length for a quantity is proportional to its diffusion coefficient. The great importance of temperature in (2.2) dictates locating the reaction-zone boundary (∗) at a fixed point on the temperature profile. According to Fig. 2.2c, the average fuel concentration in the reaction zone is then proportional to the gradient of the fuel concentration there, that is, inversely proportional to the characteristic length for fuel, or proportional to L_F. Consequently, in (2.10) and (2.12), use of $w = w_{o\infty}(L_F/\beta)^m$ is more precise than the result obtained in the previous paragraph.

Figure 2.2d illustrates the influences of varying the Lewis number for the oxidizer on the oxidizer-concentration profile. The reasoning about this effect is exactly the same as that just given. For stoichiometric mixtures, $Y_{O\infty} = 0$, and the reasoning that has been applied to the fuel also applies to the oxidizer, giving $w = w_{o\infty}(L_F/\beta)^m(L_O/\beta)^n$ in (2.10) and (2.12). For fuel-lean mixtures, $Y_{O\infty} > 0$, and it is seen that in the reaction zone Y_O is approximately constant and equal to $Y_{O\infty}$, so that $w = w_{o\infty}(L_F/\beta)^m(Y_{O\infty}/Y_{Oo})$ in (2.10) and (2.12). These results reproduce all of the major dependences of the burning velocity on properties of the combustible mixture. For example, with $m = 1$ and $n = 0$, (2.12) gives

$$v_o = \sqrt{L_F w_{o\infty} \lambda/(c_p \beta^2)}/\rho_o, \qquad (2.15)$$

which is the same as (2.7) if $\tau_c \equiv \beta^2 \rho_o/w_{o\infty}$ and $\bar{D} \equiv [\lambda/(c_p \rho_o)]^2/D_F$, in agreement with the result quoted after (2.7). The reason for the inverse dependence of the burning velocity on the diffusion coefficient is now clear—increasing the diffusion coefficient decreases the reaction rate by increasing the diffusive loss of reactant from the reaction zone, thereby decreasing the reactant concentration in the reaction zone.

The reasoning given here produces numerical corrections to the Mikhel'son formulas that can be quite substantial if overlooked. For example, if the factor β^2 in the preceding formula for τ_c were ignored, so that w is estimated as $w_{o\infty}$ in (2.6), then the predicted burning velocity would be too large by about a factor of 10; one of the β factors is associated with the small width of the reaction zone and the other with the diffusively reduced fuel concentration there. The asymptotic analysis of AEA, discussed in the next section, automatically gives these important β factors, as well as an additional numerical correction factor, typically $\sqrt{2}$. To obtain this last factor it is necessary to solve the differential equations; the simplified reasoning cannot give the constant factor. Nevertheless, it is remarkable that, without solving differential equations, burning velocities can be obtained, not only with respect to all of their functional dependences, but also with numerical accuracy of nearly 20%.

If the dependence of A on pressure is approximated by the power law $A \sim p^{\ell}$, where ℓ is the pressure exponent of the reaction rate ($\ell = 2$ for purely bimolecular reactions), then use of (2.2) in (2.12) with $w_{o\infty}$ gives

$$v_o \sim Y_{Fo}^{m/2} Y_{Oo}^{n/2} p^{\ell/2-1} e^{-E/(2R_o T_\infty)}, \tag{2.16}$$

which expresses the dependence of the burning velocity on pressure, temperature, and composition. Experimental burning-velocity dependences are often used in (2.16) to determine empirical exponents ℓ, m, and n and an empirical overall activation energy E. The -1 in $\ell/2 - 1$ comes from ρ_o in the denominator through the equation of state (1.5). Since h_o in (2.1) depends on the mixture composition (for example, moving away from conditions of stoichiometry reduces h_o approximately proportionally), with T_o fixed, T_∞ in (2.16) will vary with Y_{Fo} and Y_{Oo}; experiments with varying T_o and fixed T_∞ must be considered for extracting exponents m and n. Values of m and n usually are positive and between 0 and 1. Values of ℓ typically lie between 1 and 3, so that v_o may decrease or increase with increasing p, although $\rho_o v_o$ increases. With concentrations and pressure fixed, a variation of T_o produces only a variation of T_∞, and a graph of $\ln v_o$ as a function of $1/T_\infty$, the famous Arrhenius plot, gives E from its slope, which is $-E/(2R_o)$. The factor $1/2$ here arises from the square-root dependence of v_o on w, which is characteristic of flame propagation, as seen in (2.6).

Asymptotic Analysis of Flame Structure

A boundary-value problem for a system of ordinary differential equations with an eigenvalue (v_o) to be determined can be posed from (1.1) through (1.4) for calculating flame structures and burning velocities. For illustrative purposes the reaction $F \to P$ may be considered, with $N = 2$ and $L = 1$, so that the subscript k can be omitted in (1.4). Then $\nu_F = -1$, $W_F = W_P$, and $X_F = Y_F$. Attention is restricted to steady flow in the x direction, so that the transient-accumulation parameter

$\omega = 0$, and (1.1) gives $\rho v =$ constant. In the limit of zero Mach number, $M = 0$, (1.2) gives $p =$ constant. Considering $\boldsymbol{f}_i = \boldsymbol{f}_j$ (for example, only gravitational forces), $\alpha_{ij} = 0$ and $d_{ij} = 1/\rho$, (1.7) reduces to (1.8), and with ℓ selected so that $R = 1$, (1.8) is simply $\rho S Y_F V_F = -dY_F/dx$. Let us investigate systems with $S = 1$, f_k constant and $K_k = \infty$, so that (1.4) becomes

$$\rho v d Y_F/dx = d^2 Y_F/dx^2 - A_1 Y_F^n e^{-\beta_1/T}, \qquad (2.17)$$

where $A_1 \equiv D_1 f_1$ and $n \equiv n_{F1}$. Assuming, further, $h_1 = 1$, $h_2 = 0$, $c_p = 1$, $\lambda = 1$, $P = 1$, and $B = \infty$, we find that (1.3) is

$$\rho v dT/dx = d^2 T/dx^2 + \alpha A_1 Y_F^n e^{-\beta_1/T}, \qquad (2.18)$$

where use has been made of (2.17). It is convenient to let the reference temperature be the adiabatic flame temperature, so that α is the temperature change divided by the final temperature. Boundary conditions for (2.17) and (2.18) are then $Y_F = Y_{Fo}$ and $T = 1 - \alpha$ at $x = -\infty$ and $Y_F = 0$, $T = 1$ at $x = \infty$.

Multiplication of (2.17) by α and addition to (2.18) gives an equation relating T to Y_F, which has the solution $T = 1 - \alpha(Y_F/Y_{Fo})$. Equation (2.17) then yields a single differential equation, which can be written as

$$d^2 Y/d\xi^2 - dY/d\xi = \Lambda Y^n e^{-\beta Y/(1-\alpha Y)}, \qquad (2.19)$$

subject to $Y(-\infty) = 1$, $Y(\infty) = 0$, where $Y \equiv Y_F/Y_{Fo}$, $\xi \equiv \rho v x$, and

$$\Lambda \equiv A_1 e^{-\beta_1} Y_{Fo}^{n-1}/(\rho v)^2, \qquad (2.20)$$

with the Zel'dovich number β defined by (2.8). The objective is to find from (2.19) the structure, $Y(\xi)$, and the burning-rate eigenvalue, Λ, which is inversely proportional to the square of the burning velocity.

In the limit $\beta \to \infty$, it is seen from (2.19) that the reaction term is negligible unless Y is near zero. With the reaction zone placed at $\xi = 0$, the solution to (2.19) for $\xi < 0$ therefore is $Y = 1 - e^\xi$. For $\xi > 0$ the solution is $Y = 1$, and near $\xi = 0$ the reaction zone forms a boundary layer. The method of matched asymptotic expansions therefore is employed. Appropriate stretched variables for the reaction zone are $y = \beta Y$ and $\eta = \beta \xi$, in terms of which (2.19) is

$$d^2 y/d\eta^2 - \beta^{-1} dy/d\eta = \Lambda \beta^{-(n+1)} y^n e^{-y/(1-\beta^{-1}\alpha y)}. \qquad (2.21)$$

The matching conditions for (2.21) are $dy/d\eta \to -1$ as $\eta \to -\infty$ and $y \to 0$ as $\eta \to \infty$. With the expansions $y = y_0 + \beta^{-1} y_1 + \cdots$ and

$$\Lambda = \beta^{n+1}(\Lambda_0 + \beta^{-1}\Lambda_1 + \cdots), \qquad (2.22)$$

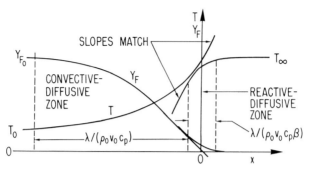

Fig. 2.3. Illustration of zone structure and matching for premixed laminar flames according to AEA.

a sequence of problems is obtained for (y_0, Λ_0), (y_1, Λ_1), $\cdots$. The first of these, found by letting $\beta^{-1} = 0$ in (2.21), is

$$d^2 y_0/d\eta^2 = \Lambda_0 y_0^n e^{-y_0}, \quad y_0'(-\infty) = -1, \quad y_0(\infty) = 0, \qquad (2.23)$$

which can be written as $\tfrac{1}{2} d(y_0')^2/dy_0 = \Lambda_0 y_0^n e^{-y_0}$, (where $y' \equiv dy/d\eta$), the integral of which, subject to the boundary conditions, gives

$$\Lambda_0 = \left[2 \int_0^\infty y_0^n e^{-y_0} dy_0 \right]^{-1}. \qquad (2.24)$$

It is possible to determine Λ_1, $\cdots$, in a similar manner. The solutions in each zone and their matching criteria are illustrated in Fig. 2.3.

The procedure that has been outlined here can be completed for more general situations. For example, for the reaction $\nu_F F + \nu_O O \rightarrow$ products, with the rate expression (2.2), it is found that, in the first approximation, the burning velocity is

$$v_o = \left[\frac{2\lambda_\infty (\nu_O W_O/\nu_F W_F)^n \rho_\infty A_\infty Y_{Fo}^{m+n-1}}{c_{p\infty} \rho_0^{m+n+2} (\bar{W}_o/W_F) \beta^{m+n+1} L_F^{-m} L_O^{-n}} \right]^{1/2} \sqrt{G} e^{-E/(2R_o T\infty)}, \qquad (2.25)$$

where the average molecular weight is $\bar{W} = \sum_{i=1}^{N} X_i W_i$, and

$$G \equiv \int_0^\infty y_0^m (y_0 + a)^n e^{-y_0} dy_0, \qquad (2.26)$$

in which $a = \beta(1 - \phi)/(L_O\phi)$, the fuel-air equivalence ratio ϕ being taken here to obey $\phi \leq 1$. Equation (2.25) allows for fully variable properties and both stoichiometric ($a = 0$) and off-stoichiometric

$(a \neq 0)$ conditions. By taking $a = 0$ and $a = \infty$, the results summarized in the previous section are recovered.

An important aspect of the AEA approach summarized here is that if accuracy in v_o comparable with that obtained experimentally and by numerical integrations (a few percent) is wanted, then two terms in the expansion should be retained (since $\beta \sim 10$). Two-term expansions, with general transport equations, are now available in the literature (for example, Chelliah and Williams, 1987). Use of these results enables v_o to be obtained from asymptotics, for comparison with experimental values and with results of numerical integrations.

However, there is a peculiarity in the problem addressed here. The problem defined by (2.19) and its boundary conditions is ill-posed in that the boundary condition $Y(-\infty) = 1$ cannot be satisfied. Inspection of (2.19) shows that in the fresh mixture ($Y = 1$), the reaction term becomes $\Lambda e^{-\beta/(1-\alpha)}$, which is small when β is large, but not zero. Since the reaction rate does not vanish, it is impossible for the fuel concentration to approach a constant value at an infinite distance ahead of the flame. At any finite burning velocity, the fuel will have reacted before reaching us. Therefore the burning velocity wants to be infinite to make Λ zero, so that the reaction term vanishes, but this clearly does not describe the real flame (and cannot satisfy the burnt-gas boundary condition). The ill-posed nature of the problem has been termed the cold-boundary difficulty.

The cold-boundary difficulty is associated with the non-vanishing of the Arrhenius rate in the reactant mixture and has been remedied in various ways, for example by placing a "flame holder" at a finite location and by applying there the boundary conditions that $Y = 1$ and that the temperature gradient take on a finite value measuring the rate of heat loss to the flame holder. If the calculated burning velocity is plotted as a function of the rate of heat loss to the flame holder, then it is found to be infinite for a zero loss rate and zero for an infinite loss rate, but to achieve a nearly constant "pseudostationary" value over a wide range of physically reasonable, intermediate loss rates. The pseudostationary value is near that found by the asymptotic expansion. The difficulty is in the physics, not in the mathematics—we do not know how best to formulate a well-posed problem. The implication is that, in fact, in the laboratory, the burning velocity is not defined precisely, but rather its value will depend somewhat on the particular experiment. The asymptotic approach captures the essence of the flame structure and provides burning-velocity results with inaccuracies comparable with differences in values obtained from different experiments. In this sense, the concept of the existence of a burning velocity fundamentally is an asymptotic concept.

A physical approach to AEA is given by Zel'dovich and Frank-Kamenetskii (1938), and a more formal approach by Bush and Fendell (1970); more recent books addressing the subject include those by Zel'dovich et al. (1980) and by Williams (1985).

Flame Instabilities

Steady, planar flow was hypothesized in the previous section. It will be seen in the next section that often time-dependent conservation equations are employed in calculating flame structures and burning velocities by numerical integration. Consideration of time-dependent or nonplanar configurations raises questions of the stability of the flame structure just described. Do small perturbations (always present in reality) to the steady, planar flame calculated here tend to grow? If so, into what structures do they evolve? These stability questions have many ramifications in combustion.

Combustion instabilities can be classified (Barrère and Williams, 1969) as intrinsic instabilities (which are associated with the combustion process itself and present even if there are no surroundings with which the combustion can interact), chamber instabilities (which, for combustion occurring within a combustion chamber, are associated with interaction of the combustion with the chamber), and system instabilities (which, in combustion systems having a combustion chamber, an intake, an exhaust, etc., involve interactions of combustion-chamber processes with processes in one or more of the other parts of the system). A wide range of physical mechanisms may be involved in these combustion instabilities, as listed in Table 2.1. The first three listed arise from interactions with small-amplitude and large-amplitude pressure waves and with the fluid flow; they occur in chamber and system instabilities. The last four can be intrinsic to the combustion process. When

Table 2.1. Mechanisms of combustion instabilities

---
1. Acoustic
2. Shock-wave
3. Vortex-shedding
4. Chemical-kinetic
5. Diffusive-thermal
6. Hydrodynamic
7. Buoyancy or acceleration

---

certain combustible mixtures are admitted into appropriate heated vessels, sequences of colorful reaction fronts are observed to move repeatedly through the vessel; these "cool flames" (an example of which may be the "will-o'-the-wisp," blue light sometimes seen in swamps at night and possibly resulting from the chemistry of methane oxidation) arise from an oscillatory character of the chemical kinetics of oxidation (possibly also involving thermal effects) and are members of a wide range of known chemical-kinetic instabilities. At a larger scale, the heat-conduction and molecular-diffusion processes in flames can interact to produce an intrinsic diffusive-thermal instability. At a somewhat larger scale than this, hydrodynamic effects associated with the density decrease across a flame give rise to an intrinsic hydrodynamic instability, discovered independently in theoretical analyses by Darrieus (1938) and Landau (1944). At still larger scales, buoyancy or acceleration of flames can produce body-force instabilities, intrinsic instabilities that arise from body-force terms in the conservation laws. Understanding why stable premixed flames can be observed in the laboratory entails comprehension of the interplay among these last three kinds of instability.

   In mathematical investigations of flame instabilities, (1.1) through (1.7) are employed with the variables appearing therein set equal to the sum of the known steady solution and a perturbation dependent on time or, more generally, both space and time (see, for example, Clavin, 1985). The stability characteristics are then determined by solving the equations for the perturbations. Underlying physical concepts, however, enable the nature of the results to be understood without fully pursuing the mathematical analysis.

   For flames propagating upward in the laboratory, the product gas below is much less dense than the reactant mixture above, and therefore the downward force of gravity is larger for a gas element above than for a gas element of the same volume below. A perturbation of the planar flame therefore will tend to result in turnover of the gas, with the fresh mixture moving downward with respect to the burnt gas, just as in unstable stratification in oceans or in an experiment in which a glass of water is turned upside down. This body-force instability causes flames propagating upward in tubes to assume a finger-like shape, the same as a large bubble moving upward in a capillary tube initially filled with water. A buoyant velocity can be estimated from the acceleration of gravity g and a characteristic length ℓ, for example the tube diameter, as $v_b = \sqrt{g\ell}$. This is approximately the velocity at which a bubble rises up a capillary tube, and if it is large compared with the laminar burning velocity v_o, then it is also approximately the velocity at which the

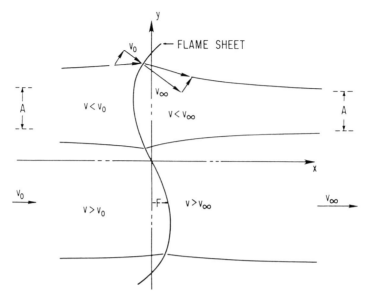

Fig. 2.4. Schematic illustration of the mechanism of hydrodynamic instability.

finger-shaped flame propagates up the tube in the combustion experiment. A characteristic wavelength for this body-force instability of a laminar flame is v_o^2/g, and its characteristic growth time is of the order of v_o/g. Thus upward-propagating planar flames are unstable, but downward-propagating planar flames tend to be stabilized by this body-force effect. In steady flows, there is a corresponding stabilizing influence for gases moving upward through a stationary flame; this contributes to stability of Bunsen flames and of flames on flat-flame burners.

The mechanism of hydrodynamic instability can be understood by reference to Fig. 2.4. Since the density ρ_o of the fresh mixture exceeds the density ρ_∞ of the burnt gas, the velocity v_o of the fresh mixture is less than the velocity v_∞ of the burnt gas for steady, planar flames by mass conservation (1.1) (recall $\rho v =$ constant). If the planar flame is perturbed, then continuity of the component of velocity tangential to the flame implies that the streamlines must diverge ahead of a bulge protruding into the fresh mixture and converge behind this bulge, as illustrated in Fig. 2.4. Since mass conservation in a variable-area flow states that $\rho v A =$ constant, where A is the cross-sectional area, it is seen that just ahead of this bulge the velocity will be less than v_o (and just behind it the velocity will be less than v_∞). Therefore locally the bulge propagates faster than the local fluid velocity, and the bulge tends to move even farther upstream. Thus, the perturbation will tend to grow by this purely hydrodynamic mechanism. If v_∞ had been less than v_o, then the jump conditions across the flame would show that the stream-lines converge rather than diverge ahead of the bulge, and the flame would have been found to be stable. Since $v_\infty > v_o$ for all real flames, they all are unstable to this hydrodynamic mechanism. In view of lab-oratory observations of stable, planar, laminar flames, publication of their theoretical predictions required courage on the part of Darrieus and Landau. The stability problem can be addressed mathematically by looking for solutions to (1.1) through (1.4) in which the variables are set equal to those obtained from the steady, planar solutions plus a small constant times a function of x times $e^{\sigma t} \sin(ky)$, where y is a transverse coordinate, k is a constant wave number (reciprocal of a wavelength), and σ is a constant growth rate (reciprocal of a growth time). When this is done, a dispersion relation expressing σ as a function of k is obtained, as in numerous problems in other branches of physics. With gravitational acceleration g included as a stabilizing influence, the dis-persion relation is found to be

$$\sigma = [v_o k/(1+\rho)]\left\{\sqrt{1 + [(1-\rho^2)/\rho][1 - g/(kv_o v_\infty)]} - 1\right\}, \quad (2.27)$$

where the density ratio is $\rho = \rho_\infty/\rho_o = v_o/v_\infty < 1$. This result shows that for long wavelengths (small k), that is, for $k < g/(v_o v_\infty)$, buoyancy stabilizes the hydrodynamic instability when $g > 0$. However, at smaller wavelengths the hydrodynamic instability always prevails.

Only diffusive-thermal effects can stabilize the hydrodynamic insta-bility at small wavelengths and enable steady, planar flames to be seen in the laboratory. Influences of diffusive-thermal effects can be under-stood by considering the flame structure of Fig. 2.3. The thin reaction zone can be approximated as a reaction sheet, and effects on the flame structure caused by perturbing this sheet from a planar configuration can be investigated, as illustrated in Fig. 2.5. The reaction sheet is a source for heat and a sink for reactants. Ahead of the sheet, conduction of heat and diffusion of reactants occur, as explained in the previous sections. If the diffusivities of reactants and of heat are the same (a Lewis number L of unity, "normally diffusing reactants" in Fig. 2.5), then when there is a bulge in the sheet toward the fresh mixture, the en-hanced conductive heat loss from it is balanced by the enhanced rate of diffusion of reactants into the reaction zone, so that an energy balance for the reaction sheet shows no change in its temperature (the flame temperature, T_f). For weakly diffusing reactants $(L > 1)$, heat is lost by conduction more readily from the sheet than it can be generated by diffusion of the reactant, and T_f is decreased at the forward-pointing bulge. For strongly diffusing reactants $(L < 1)$, conversely T_f will be

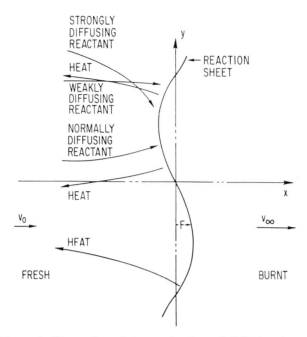

Fig. 2.5. Schematic illustration of the mechanism of diffusive-thermal instability.

higher at the bulge. As (2.2) and (2.6) imply, an increase in T_f results in an increased burning velocity and a tendency for the bulge therefore to propagate farther upstream. Therefore the planar flame is reasoned to be unstable by a diffusive-thermal instability for $L < 1$ and stable for $L \gtrsim 1$. Flames with $L > 1$ thus may have diffusive-thermal stabilizing influences at sufficiently small wavelengths (comparable in size with the flame thickness δ) that can offset the hydrodynamic instability, rendering downward-propagating, planar flames stable.

This reasoning is only qualitative, and mathematical analysis of (1.1) through (1.4) is needed for investigating the stability of steady, planar flames with diffusive-thermal effects included. A number of analyses have been completed, and various dispersion relations are available. One set of results (Joulin and Clavin, 1979) is shown in Fig. 2.6, where $\rho = 1$ has been introduced to focus on diffusive-thermal effects by removing hydrodynamic and body-force instabilities, and where radiant heat loss (through an approximation in (1.3) for $\nabla \cdot \boldsymbol{q}/B$) has been included. In this figure μ is the ratio of the burning velocity of the planar flame with heat loss to that of the adiabatic planar flame, and therefore $1/\mu$ is a measure of the heat loss. For the adiabatic flame $\ln(1/\mu^2) = 0$, and the nonplanar instability just described is found to occur for $L < 1 - 2/\beta$,

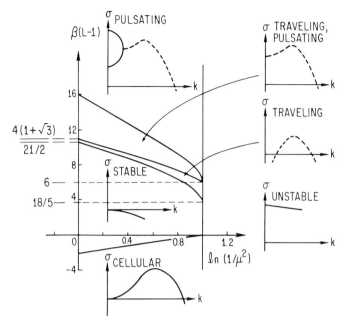

Fig. 2.6. Regions of diffusive-thermal instability in a Lewis-number, heat-loss plane, with dispersion relations illustrated by insets.

the Zel'dovich number β being given in (2.8). Since the reasoning in Fig. 2.5 neglected the thickness of the reaction zone, which is $1/\beta$ times the flame thickness, it is understandable that the critical Lewis number obtained there is wrong by an amount of order $1/\beta$.

Figure 2.6 shows the existence of other types of diffusive-thermal instabilities, as yet unanticipated. The diffusive-thermal effects for the adiabatic flame are found to be stabilizing only if $1 - 2/\beta < L < 1 + 21\beta/2$. The characters of the various instabilities are shown by the dispersion-relation diagrams in Fig. 2.6, where dashed curves identify traveling-wave instabilities (for which σ has a nonzero imaginary part). When L is large enough, there is a new mechanism of instability, corresponding to a pulsating, planar flame; a physical description of the mechanism of this instability also can be given. The pulsating instability is observed mainly for planar deflagration of solid combustibles because molecular diffusion coefficients of solids are small enough compared with their thermal diffusivities for their Lewis numbers to be large. The instabilities for $L < 1 - 2/\beta$ are identified in Fig. 2.6 as cellular instabilities because the mechanism of Fig. 2.5 leads to flames having nonplanar, cellular-like structures (cellular flames). The range of stability is seen in Fig. 2.6 to decrease with increasing heat loss and to be quite narrow at the critical value of heat loss needed to extinguish the flame. Therefore, under appropriate conditions diffusive-thermal instabilities lead to many interesting flame behaviors, such as rotating polyhedral flames and spinning deflagrations (most often observed for solid fuels with solid products).

When density changes are taken into account, the range of stabilizing diffusive-thermal effects is found to widen, in comparison to Fig. 2.6. The gas expansion lengthens distances, decreasing gradients and therefore fluxes into the reaction zone in Fig. 2.5. The instability mechanism of Fig. 2.5 therefore tends to be suppressed by the gas expansion. A result (if the thermal conductivity is constant) for adiabatic flames is that the cellular instability occurs only if

$$L < 1 - [2 \ln(1/\rho)] \left[\rho\beta \int_0^{(1-\rho)/\rho} \ln(1+z)dz/z \right]^{-1}, \qquad (2.28)$$

which reduces to the constant-density result for $\rho = 1$, gives $L < 1 - 3.37/\beta$ for $\rho = 1/2$, and $L < 0.5$ for $\rho = 1/4$, $\beta = 10$. Of all common flames, only fuel-lean hydrogen flames have $L < 0.5$. Therefore the diffusive-thermal effects are indeed stabilizing for most real flames, and the planar structures that are observed are expected. Lean hydrogen-air flames experimentally exhibit strong cellular instability; mixtures ignited at the top of a tube develop pronounced fingers of flame that propagate downward, leaving behind unburnt fuel between the fingers. This is associated with the high diffusion coefficient of hydrogen, which readily diffuses preferentially to the tips of the fingers, intensifying the burning (and increasing the burning velocity) there while leaving behind a mixture too lean to burn. The cellular structure of lean hydrogen flames has a bearing on safety hazards in nuclear reactors—reactions of hot fuel cladding with water liberate hydrogen into the air of the containment vessel, forming a mixture that can then ignite and burn, producing high overpressure, as occurred at Three Mile Island and Chernobyl. Thus it is important to know how lean a mixture must be to prevent combustion (see Chapter 4).

A great deal of fundamental information on flame instabilities was obtained by Markstein (1964), and a thorough recent review is provided by Clavin (1985).

Influences of Kinetic Mechanisms

According to the simplest ideas of chemical kinetics thus far employed, the burning velocities shown in Fig. 2.1 should attain maximum values at stoichiometry (an equivalence ratio of unity) because the adiabatic flame temperature (and hence the reaction rate in the reaction zone) is greatest there. Experimentally the maxima are seen to occur at somewhat fuel-rich conditions. Some reasons for such shifts from stoichiometry can be found in the ideas presented previously. The burning velocity was seen to decrease with an increasing diffusion coefficient of the fuel for fuel-lean mixtures. Similarly, it should decrease with an increasing diffusion coefficient of the oxidizer for fuel-rich mixtures, and be nearly independent of the diffusion coefficient of the fuel under these rich conditions. These transport effects thus should cause fuels having high diffusion coefficients to tend to exhibit fuel-rich burning-velocity maxima. Hydrogen is such a fuel, and in fact hydrogen-air flames exhibit a fuel-rich burning-velocity maximum largely for this reason. However, most of the fuels of Fig. 2.1 are large molecules with low diffusion coefficients, and therefore the transport effects should tend to produce fuel-lean burning-velocity maxima for them. The maxima on the rich side are a result of changes in chemical kinetics with the change in equivalence ratio; equation (2.2) cannot be used with the same constant parameters A, m, n, and E for both rich and lean mixtures. Attention must be given to kinetic mechanisms if fundamental explanations of these effects are to be obtained.

Three different approaches can be identified for obtaining influences of the real chemical kinetics on flame structures and burning velocities. These three approaches are (1) experiments on laminar flames, (2) numerical integration of the conservation equations with complete chemical kinetics included, and (3) asymptotic methods generally involving steady-state approximations for reaction intermediaries. These methods will be discussed in the next few pages.

The first approach is through experiments on laminar flames. If profiles of temperature and of chemical composition through a flame can be measured, then much information about kinetic mechanisms can be extracted. Experiments of this kind are difficult. They require the flame to be held quite steady in the laboratory for the measurement duration, and they rely on the development of instrumentation capable of fine enough spatial resolution to obtain profiles. Until recently this was practically impossible for most flames at atmospheric pressure because their thicknesses (< 1 mm) are too small. Extensive experimentation therefore was performed at reduced pressures where, according to (2.2) with $A \sim p^{\ell}$ and (2.4), the thicknesses are larger (in proportion to $p^{-\ell/2}$). Flames 1 cm or more thick were established, and temperature profiles were measured by thermocouples, while gas samples were withdrawn by fine sampling probes, and the samples were analyzed by mass spectrometry to identify the molecules present and by gas chromatography to measure their concentrations quantitatively. Although sampling techniques were studied carefully, advanced, and perfected to be sure that the sample closely represented the gas actually present in the flame, it remained largely impractical to obtain good quantitative measurements of unstable intermediate species such as free radicals (OH, CH, etc.), which are very important to the chemical kinetics. Profiles from sampling were restricted mainly to stable major species and to low-pressure flames. Fine thermocouples, oriented parallel to the flame sheet to minimize conductive losses and suitably corrected for radiant energy losses and for surface catalytic effects, were more easily developed to provide good data in flames at pressures up to atmospheric.

In recent years laser diagnostic techniques have advanced to a point at which thcy give good data on flame structure even at atmospheric pressure, with spatial resolutions approaching 10μ. The most useful method has involved laser Raman spectra which, through a great deal of work in laser development and in characterization of the Raman spectra of different molecules, has succeeded in measuring profiles of most stable species and many radicals, as well as temperature (usually through its effect on the N_2 spectrum). Detailed experimental information can now be obtained on flame structures by these various techniques. This information helps greatly in identifying the chemical-kinetic mechanisms of flames.

A second approach is through numerical integration of the conservation equations with complete chemical kinetics included. For this purpose (1.1) through (1.4) may be reduced to a system of ordinary differential equations for describing steady, planar flame structure. A convenient dependent variable is the mass flux fraction of species i, $\epsilon_i = Y_i(1 + V_i/v)$, since changes in this appear directly in the species conservation (1.4). In dimensional variables, the system

$$\lambda dT/dx = \rho_o v_o \sum_{i=1}^{N} (h_i \epsilon_i - h_{io}\epsilon_{io}) , \tag{2.29}$$

$$\rho_o v_o d\epsilon_i/dx = W_i \sum_{k=1}^{L} \nu_{ik} k_k \prod_{j=1}^{N} [(X_j p)/(R_o T)]^{n_{jk}} , \tag{2.30}$$

$$dX_i/dx = (\rho_o v_o/\rho) \sum_{j=1}^{N} (X_i X_j/D_{ij})(\epsilon_j/Y_j - \epsilon_i/Y_i) \tag{2.31}$$

is obtained, where thermal diffusion has been neglected in (1.7), forward and backward directions of the same step are now counted separately (increasing L), and the specific reaction-rate constant for the k'th step is approximated as

$$k_k = B_k T^{\alpha_k} e^{-E_k/(R_o T)}, \qquad (2.32)$$

with B_k, α_k, and E_k constants. The orders n_{jk} now are molecularities that represent the number of molecules that collide when the reaction occurs. Given the ideal-gas equation of state (1.5) and the formula for Y_i in terms of X_i, (2.29) through (2.31) comprise $2N - 1$ independent equations for the $2N - 1$ variables T, ϵ_i, X_i, $i = 1, \ldots, N - 1$. Along with suitable boundary conditions at $x = \pm\infty$, it apparently should be possible to integrate the system of equations numerically and to determine the value of the burning velocity v_o for which the boundary conditions can be satisfied.

This plan for numerical integration can be carried out only if the thermodynamic properties (h_i), the transport coefficients (λ, D_{ij}), and the chemical-kinetic constants $(n_{jk}, B_k, \alpha_k, E_k)$ are known, as well as the stoichiometry (ν_{ik}) of all reactions that might occur. In principle there is always a question as to whether all of the potentially relevant elementary reaction steps have been discovered, but even more importantly, there are uncertainties in all of the other aspects just cited. Experiments play significant roles in reducing the uncertainties. Independent thermodynamic measurements underlie determinations of h_i. Independent transport measurements underlie determinations of λ and D_{ij}. Finally, independent measurements of rates of individual chemical reaction steps, for example by use of shock tubes or of molecular beams, underlie determinations of the chemical-kinetic constants. In all cases, principles of the theory in the subject aid in selecting the best values, but the independent experiments nevertheless are of great importance. In principle, these experiments are not performed with flames, so that the predictions through numerical integrations are not based on the structures that they are trying to predict. In practice, there are uncertainties, especially in some elementary rate constants, B_k, α_k, and E_k, and (usually to a lesser extent) in some transport coefficients. Values of the most uncertain constants therefore often are adjusted to improve agreement with flame experiments.

Just as has been indicated earlier for the simple one-step chemistry, the problem defined here for numerical integration is ill-posed as a consequence of the cold-boundary difficulty. Nevertheless, a "flame holder" can be introduced, or all rates may arbitrarily be put equal to zero at temperatures below a value perhaps 100 K above the initial temperature, to achieve a well-posed problem. The integration can then be performed, at least in principle, and a pseudo-stationary value of v_o can be sought, if desired. In fact, with a large number of species and reactions, the numerical integration is difficult because of singularities at the end points. Therefore, nowadays practically none of the approaches to numerical integration employ (2.29) through (2.31). Instead, they use one-dimensional, time-dependent versions of (1.1) through (1.4) and march forward in time until the steady structure is attained. There are methods, for example iterative methods, that appear to employ (2.29) through(2.31), but that in fact are more nearly equivalent to marching in time (for example, in the course of the iteration). With the forward marching, the problem is parabolic, and the solutions usually remain well-behaved. The penalty paid is that a problem in one dimension (x) has been converted to one in two dimensions (x, t), thereby reducing the number of computation points in x that can be retained within a fixed computational capacity. The numerical solution is less accurate than it would have been if the marching in time were unnecessary. Thirty years ago the accuracies achieved were quite poor, but computer speed and capacity has increased so much that numerical integrations now give flame structures with accuracies equal to or better than those attained experimentally.

What we have learned in the preceding section about instabilities has a bearing on what can be expected from these numerical integrations. If the flame possesses a pulsating instability, the time-marching integration should reveal it and not settle down to a steady structure. This has the advantage of demonstrating the instability, but the disadvantage is that the steady structure cannot be obtained (for many purposes, for example in stability studies, it is desirable to know what the steady structure is, even if it is unstable). On the other hand, if the flame possesses a cellular instability, the time-marching integration cannot reveal it. Two-dimensional or three-dimensional time-dependent equations would have to be integrated numerically to see cellular instability, and with complete hydrocarbon-air combustion chemistry, for example, this will remain beyond supercomputer capacities for many years to come. The current methods thus have the disadvantage of not showing cellular structure, but the advantage of being able to give the steady, planar structure even if the flame is cellularly unstable (a full three-dimensional code would not do this, although presumably it could readily be restricted to one dimension to do so). For investigating nonplanar, time-dependent flames or flames in nonuniform flows within current computational capabilities, simplifications to complete chemistry are needed. These simplified descriptions have been found to be achievable through asymptotic methods—the third type of approach to studying structures of flames with real chemistry.

Activation-energy asymptotics, described earlier for the one-step approximation, in principle can similarly be applied to each elementary step that has an E_k large enough. Most unimolecular and two-body steps have sufficiently large E_k's, but many three-body processes, especially radical combinations (such as $H_2 + O + M \rightarrow H_2O + M$, where M represents any third body), have $E_k = 0$ and therefore are not amenable to AEA. A different kind of asymptotics, rate-ratio asymptotics (RRA), has now been found to play a greater role than AEA for flames with sufficiently complicated kinetics. The basis of RRA, which was identified briefly in Chapter 1, can be discussed by considering the steady-state approximation for reaction intermediaries, an approximation that has been employed in chemical kinetics for many years.

In (2.30), usually some of the reaction terms are positive and others negative. If w_{i+} denotes the sum of all positive terms and $-w_{i-}$ the sum of all negative terms, then (2.30) can be written as

$$\rho_o v_o d\epsilon_i/dx = w_{i+} - w_{i-} . \qquad (2.33)$$

The steady-state approximation is to assume that w_{i+} and w_{i-} are both large compared with $\rho v d\epsilon_i/dx$, so that (2.33) can be approximated as

$$w_{i+} = w_{i-} . \qquad (2.34)$$

This approximation can be derived formally by taking the limit of large values of all Damköhler numbers D_k in (1.4), and therefore RRA often is equivalent to Damköhler-number asymptotics (DNA) or the fast-chemistry limit. This limit is singular in that the differential equation (2.33) is replaced by the algebraic equation (2.34), so that the order of the differential system is reduced. The number of permissible boundary conditions therefore is decreased, but in flame propagation often the original conditions turn out to be closely consistent with the reduced set in the reaction regions, so that no significant inconsistencies arise. The simplification to the computational problem afforded by (2.34) is that the number of differential equations that need to be integrated is reduced. The mole fraction X_i of the species to which the steady state is applied is eliminated from the system by solving (2.34) for X_i in terms of the other variables. Completion of the integration with the simplified kinetics then gives $X_i(x)$ as well, and from this and the rest of the solution, $\rho_o v_o d\epsilon_i/dx$ can be calculated and compared with w_{i+} to test the accuracy of the approximation. If the approximation is deemed to be accurate enough, then a simplified description of the flame structure has been obtained that reduces computational tasks (which can be especially useful in more complicated problems that involve nonuniform flows) and that enhances our comprehension of the structure.

There is a potential pitfall in application of the steady-state approximation. The combination $w_{i+} - w_{i-}$ may by chance occur in the rate expressions for other species, and in some of these the additional rate terms may not be large compared with the difference, $w_{i+} - w_{i-}$. In this case (2.34) must not be used directly in the rate equation for the other species, but instead (2.33) must be substituted there, with $d\epsilon_i/dx$ obtained through the $X_i(x)$ calculated by (2.34). Progress has been made in the development of formal mathematical procedures for circumventing this pitfall.

Each steady-state approximation that is introduced reduces the order of the differential system. If steady states for all reaction intermediaries are postulated, then one-step chemistry is recovered, but with a rate expression more complex than the simple Arrhenius formula (2.2). Nevertheless, AEA often can be applied to the resulting rate to provide simple formulas for burning velocities and flame structures. In this way, asymptotic methods involving combinations of RRA and AEA can be employed with detailed chemistry.

Often steady states cannot be justified well for all intermediaries, and instead two-step, three-step, or four-step mechanisms are derived. Analysis of the flame structure then becomes more complicated and often necessitates numerical integrations, but these usually can be performed more simply than with full chemistry. Different approximations of the same general type as steady-state approximations sometimes can be justified when steady states cannot. The most common examples are partial-equilibrium approximations. If (2.30) is revised to include forward and backward rates of each step k, as in (1.4), then with w_{kf} representing the forward rate and w_{kb} the backward rate, the partial-equilibrium approximation for step k is

$$w_{kf} = w_{kb} . \qquad (2.35)$$

This can be derived formally from (1.4) by considering the limit of large D_k for this particular k. Evidently (2.35) can be used in the same manner as (2.34) to eliminate X_i for one of the species i that enters into reaction k, and the same type of simplification achieved with the steady-state approximation is obtained. The same pitfalls arise and should be circumvented in a similar manner. The validity of the approximation is tested by evaluating $d\epsilon_i/dx$ for the eliminated i and employing the conservation equation for this species.

Steady-state and partial-equilibrium approximations are examples of a general class of RRA approximations. In principle, in the system (2.29) through (2.31) linear combinations of dependent variables can be introduced that locally diagonalize the matrix on the right-hand side, and from the eigenvalues of the diagonalized form an ordering of chemical times from fast to slow can be made. For the fast times, algebraic eliminations like those from (2.34) or (2.35) are made, and a reduced system of differential equations is then obtained to be solved. In applications, the general approach often reduces to the steady-state or partial-equilibrium methods identified above. Discussions of specific flames are of interest to illustrate the ideas that have been introduced here.

Extensive use of steady-state approximations for various flames was made in the work of von Kármán and coworkers (1954). Experimental aspects of structures of laboratory flames are discussed by Fristrom and Westenberg (1965), by Glassman (1987), and by Lewis and von Elbe (1987), for example, who also present theoretical concepts.

Examples of Real Flames

The ozone decomposition flame is the simplest one for theoretical studies because only one element (oxygen) is involved, and only three molecules, the O atom, the O_2 molecule, and ozone O_3. Experimentally this flame can be dangerous to investigate because of the high reactivity of O_3. The overall chemistry is

$$O_3 \rightarrow (3/2)O_2 , \tag{2.36}$$

which is exothermic, releasing an amount of energy here denoted by Q. Only six elementary steps are potentially relevant,

$$O_3 + M \rightarrow O_2 + O + M , \tag{2.37}$$

$$O_3 + O \rightarrow 2O_2 , \tag{2.38}$$

$$2O + M \rightarrow O_2 + M , \tag{2.39}$$

and their reverse steps. In fact, O_2 is so stable that the reverses of (2.38) and (2.39) are negligible, except for determining the final state when adiabatic flame temperatures are high enough for equilibrium dissociation to become important. The ozone dissociation (2.37) is endothermic, requiring energy denoted here by D. The ratio

$$c = D/Q \tag{2.40}$$

is about 0.7 for the ozone flame. The O atom is active, so that the step (2.38) proceeds rapidly with a small activation energy E_2. On the other hand, the dissociation (2.37) requires an appreciable activation energy E_1, so that the ratio

$$b = E_2/E_1 \tag{2.41}$$

is only about 0.2.

The O atom is so reactive that introduction of a steady-state approximation for it is attractive. When this is done, a one-step approximation for the chemistry in this flame is obtained. At the adiabatic flame temperature, over a wide range of practical conditions the rates of (2.37) and (2.38) are large compared with that of (2.39), so that in analyzing flame propagation (2.39) may be neglected in a first approximation. The recombination (2.39) occurs in a long downstream recombination zone that maintains a convection-reaction balance and that does not affect the burning velocity. Concerning propagation, therefore, there is a relevant flame temperature T_f at the beginning of the recombination zone; T_f is determined by the chemical kinetics and is less than the adiabatic flame temperature T_∞. There are other flames that also exhibit long recombination zones that have little influence on the burning velocity.

For the ozone flame the complication associated with the recombination zone is readily handled by asymptotic methods, and AEA for E_1 may be applied with the steady state to determine a structure otherwise like that illustrated in Fig. 2.3. There is a preheat zone followed by a thin reaction zone in which only (2.37) and (2.38) occur. The O-atom steady state is accurate in the reaction zone only at low initial ozone mole fractions X_{3o}; at higher X_{3o} (2.38) is not fast enough to maintain the steady state, and the two separate rates of (2.37) and (2.38) must be retained in the reaction zone (giving a two-step mechanism). Figure 2.7 compares the burning velocity obtained by AEA with that found by numerical integration. Because of experimental difficulties, sufficiently accurate experimental results are not yet available for this flame for comparison. The agreement in Fig. 2.7 is seen to be quite good, especially at low values of X_{3o}, where the steady state applies. At higher values, results of the one-term and two-term expansions in β^{-1} are seen to differ appreciably, and the idea of selecting a value $1/3$ of the way back from the two-term expansion towards the one-term expansion is seen to give good agreement. The asymptotic expansions in β^{-1} have been found to be oscillatory, so that the one-term and two-term expansions bound the burning velocity.

Another theoretically possible structure for the ozone flame is one in which mainly (2.37) occurs in the hot reaction zone, and the O atoms produced there diffuse into the preheat zone, where the exothermic step (2.38) occurs, converting the preheat zone into an exothermic zone. With this structure, a steady-state approximation is found to apply in (2.29) for energy generation in the reaction zone (a steady state for T), and the reaction zone is found to maintain a balance between all three processes—convection, reaction, and diffusion. This type of structure, involving forward diffusion of active intermediaries in an essential way,

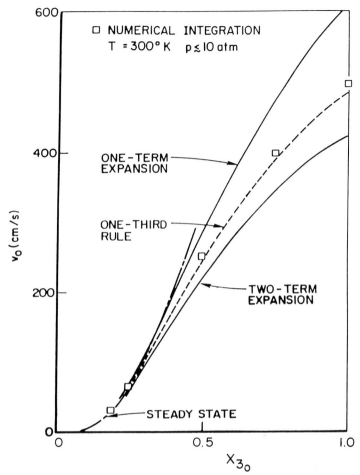

Fig. 2.7. Dependence of the burning velocity on the initial ozone mole fraction for the ozone decomposition flame.

has often been discussed as a potential mechanism for flame propagation. With the rate constants currently believed to apply in the ozone reactions, this mode of propagation never occurs for the ozone flame. Essentially, c of (2.40) is too small and b of (2.41) too large for O to penetrate far enough into the preheat zone and to release enough energy there. However, this type of structure, which has been called a two-zone structure because of the importance of two different finite reaction rates in two separate zones, might occur in other flames. Asymptotic analyses of the ozone flame appear in the work of Rogg, for example Rogg et al. (1986).

Hydrogen-halogen flames offer examples of flames having chemistry that is simpler than that of hydrocarbon-air flames but more complex than that of the ozone flame. The halogens are F, Cl, Br, and I, and a representative member of this class of flames is the hydrogen-bromine flame, the overall reaction of which is

$$H_2 + Br_2 \rightarrow 2HBr. \qquad (2.42)$$

The mechanism does not involve (2.42) because it is too slow, but rather proceeds through

$$
\begin{align}
Br_2 + M &\rightarrow 2Br + M, & (2.43) \\
Br + H_2 &\rightarrow HBr + H, & (2.44) \\
H + Br_2 &\rightarrow HBr + Br, & (2.45) \\
H + HBr &\rightarrow H_2 + Br, & (2.46) \\
2Br + M &\rightarrow Br_2 + M. & (2.47)
\end{align}
$$

Here (2.47) is the reverse of (2.43) and (2.46) the reverse of (2.44); the reverse of (2.45) seems slow enough to be always negligible. This is a chain mechanism because the reaction proceeds through active intermediaries, Br and H, which are regenerated as the reaction progresses. It is a "straight chain" because every time an intermediary is consumed another one is generated. Once a small amount of Br is formed by (2.43), it may consume all of the reactants through (2.44) and (2.45); hence the term "chain" is employed, with Br and H being the chain carriers. The ozone chemistry is not a chain reaction because the active intermediary O is not regenerated in (2.38). Step (2.43) is chain initiation, (2.44) through (2.46) are chain-carrying or chain-propagation steps because they continue the chain, and (2.47) is chain termination.

The steady-state approximation for H tends to be good for the hydrogen-bromine flame, while that for Br is of marginal accuracy. Given the steady-state for H and the mechanism (2.43) through (2.47), the steady-state for Br is formally equivalent to partial-equilibrium for (2.43), and this is in fact what has been employed in analyses, rather than a correct formulation derived solely from the two steady states. The structure of the flame bears a similarity to that of the ozone flame, with a preheat zone, a reaction zone in which (2.44) through (2.46)

Table 2.2. Values of the activation-energy and energetic parameters for the four halogen flames

Halogen	F_2	Cl_2	Br_2	I_2
b	0.04	0.12	0.45	1.11
c	0.29	1.31	1.85	15.62

occur, and recombination zones. Comparisons like those discussed for ozone have not been completed for hydrogen-halogen flames.

If the overall heat release Q is taken to be that of (2.42) and the dissociation energy D that of (2.43), then the energetic parameter c of (2.40) can be evaluated for these flames. The activation energy E_1 for the fast steps can be taken as that of (2.44), while that for dissociation E_2 is the activation energy of (2.43); the activation-energy ratio b of (2.41) therefore also can be evaluated for these flames. The results are shown in Table 2.2, where it is seen that possibly for F_2, and especially for Cl_2, b is small enough and c large enough that a two-zone structure, resembling that described in the ozone discussion, may occur under suitable conditions for these flames. More study is needed to see if this is so. For I_2, b and c are so large that a chain mechanism like (2.43) through (2.47) is unlikely; instead, a step like (2.42) may describe the mechanism in this flame (if, indeed, such a flame exists—the heat release is small, and the flame has never been seen experimentally).

Hydrocarbon-air flames are of comparatively great practical interest. These involve all of the steps of the hydrogen-oxygen flame, as well as very many steps of hydrocarbon chemistry. In these flames, just as in the hydrogen-oxygen flame, the active species H, OH, and O are important chain carriers. The representative step

$$H + O_2 \rightarrow OH + O \qquad (2.48)$$

is chain branching because two carriers are produced while only one is consumed. The mechanism therefore involves "branched-chain" chemistry, which can result in rapid build-up of concentrations of chain carriers. Many experimental results for hydrogen-oxygen and hydrocarbon-air flames are available, and numerical integrations have been employed extensively in recent years to calculate structures and burning velocities of these flames. Attention here will be focused on the methane-air flame as the simplest example of a hydrocarbon-air flame; its overall reaction was given in (1.9).

A reduced kinetic mechanism for methane-air flames appears in Table 2.3. The rate data come from careful and extensive compilations, such as those of Warnatz (1984). The first nine steps in the table are the most relevant portion of the hydrogen-oxygen chemistry. Step 10 is the mechanism by which CO is oxidized to CO_2. The remaining steps involve hydrocarbon chemistry. Important backward rates are included and identified by b. Despite this reduced numbering system, it is seen that there are 31 steps in this simplified mechanism. In fact, the mechanism is highly simplified. More than 200 steps now are routinely

Table 2.3. The elementary reaction mechanism and associated rate constants for the methane-air flame

Step	Reaction	B^*	α^*	E^*
1	$O_2 + H \rightarrow OH + O$	$2.00\ 10^{14}$	0.00	70.30
1b	$OH + O \rightarrow O_2 + H$	$1.40\ 10^{13}$	0.00	3.20
2	$O + H_2 \rightarrow H + OH$	$1.50\ 10^{7}$	2.00	31.60
2b	$H + OH \rightarrow O + H_2$	$6.73\ 10^{6}$	2.00	22.35
3	$OH + H_2 \rightarrow H + H_2O$	$1.00\ 10^{8}$	1.60	13.80
3b	$H + H_2O \rightarrow OH + H_2$	$4.62\ 10^{8}$	1.60	77.50
4	$OH + OH \rightarrow H_2O + O$	$1.50\ 10^{9}$	1.14	0.42
4b	$H_2O + O \rightarrow OH + OH$	$1.49\ 10^{10}$	1.14	71.14
5**	$H + O_2 + M \rightarrow HO_2 + M$	$2.30\ 10^{18}$	-0.80	0.00
6	$HO_2 + H \rightarrow OH + OH$	$1.50\ 10^{14}$	0.00	4.20
7	$HO_2 + H \rightarrow H_2 + O_2$	$2.50\ 10^{13}$	0.00	2.90
8	$HO_2 + H \rightarrow H_2O + O$	$3.00\ 10^{13}$	0.00	7.20
9	$HO_2 + OH \rightarrow H_2O + O_2$	$2.00\ 10^{13}$	0.00	7.20
10	$CO + OH \rightarrow CO_2 + H$	$4.40\ 10^{6}$	1.50	-3.10
10b	$CO_2 + H \rightarrow CO + OH$	$4.96\ 10^{8}$	1.50	89.71
11	$CH_4 + H \rightarrow H_2 + CH_3$	$2.20\ 10^{4}$	3.00	36.60
11b	$H_2 + CH_3 \rightarrow CH_4 + H$	$8.83\ 10^{2}$	3.00	33.53
12	$CH_4 + OH \rightarrow H_2O + CH_3$	$1.60\ 10^{6}$	2.10	10.30
13	$CH_3 + O \rightarrow CH_2O + H$	$7.00\ 10^{13}$	0.00	0.00
14	$CH_3 + OH \rightarrow CH_2O + H + H$	$9.00\ 10^{14}$	0.00	64.80
15	$CH_3 + OH \rightarrow CH_2O + H_2$	$8.00\ 10^{12}$	0.00	0.00
16***	$CH_3 + H \rightarrow CH_4$	$6.00\ 10^{16}$	-1.00	0.00
17	$CH_2O + H \rightarrow CHO + H_2$	$2.50\ 10^{13}$	0.00	16.70
18	$CH_2O + OH \rightarrow CHO + H_2O$	$3.00\ 10^{13}$	0.00	5.00
19	$CHO + H \rightarrow CO + H_2$	$2.00\ 10^{14}$	0.00	0.00
20	$CHO + OH \rightarrow CO + H_2O$	$1.00\ 10^{14}$	0.00	0.00
21	$CHO + O_2 \rightarrow CO + HO_2$	$3.00\ 10^{12}$	0.00	0.00
22**	$CHO + M \rightarrow CO + H + M$	$7.10\ 10^{14}$	0.00	70.30
23	$CH_3 + H \rightarrow CH_2 + H_2$	$1.80\ 10^{14}$	0.00	63.00
24	$CH_2 + O_2 \rightarrow CO_2 + H + H$	$6.50\ 10^{12}$	0.00	6.30
25	$CH_2 + O_2 \rightarrow CO + OH + H$	$6.50\ 10^{12}$	0.00	6.30
26	$CH_2 + H \rightarrow CH + H_2$	$4.00\ 10^{13}$	0.00	0.00
26b	$CH + H_2 \rightarrow CH_2 + H$	$2.79\ 10^{13}$	0.00	12.61
27	$CH + O_2 \rightarrow CHO + O$	$3.00\ 10^{13}$	0.00	0.00
28	$CH_3 + OH \rightarrow CH_2 + H_2O$	$1.50\ 10^{13}$	0.00	20.93
29	$CH_2 + OH \rightarrow CH_2O + H$	$2.50\ 10^{13}$	0.00	0.00
30	$CH_2 + OH \rightarrow CH + H_2O$	$4.50\ 10^{13}$	0.00	12.56
31	$CH + OH \rightarrow CHO + H$	$3.00\ 10^{13}$	0.00	0.00

* Here cm, mol, K and kJ are the units.
** Catalytic efficiencies differ for different M; values here are for $M = H_2$.
*** The high-pressure value k_∞ is given here; tail-off curves are $k/k_\infty = (1 + 21.5 \times 10^{10} T^3/p^{0.6})^{-1}$, where p is in atm and T in K.

employed in calculating hydrocarbon flame structures by numerical integration. The values of the rate constants listed follow the notation of (2.32) and comprise one particular selection; there are some significant differences in constants employed by different investigators, especially for the fuel chemistry, although the entries for the first eleven steps (the most important ones for the present problem) now rest on fairly firm ground. Some rates, such as that of step 10, are approximated better by a formula such as $k_k = B_k e^{\alpha_k T}$, rather than (2.32), if the temperature range is large, but (2.32) is adequate when attention is restricted to the temperature ranges of importance in flame propagation.

Not all steps are equally important. Many may be neglected entirely, although there always remains a concern that the addition of a large number of small terms may result in a significant contribution. To understand the structure, simplifications are needed, and these are provided mainly by steady-state and partial-equilibrium approximations. As a starting point, we may attempt to guess a minimal set that may describe the main features of the flame, and introduce steady-state approximations in this set to obtain a reduced mechanism. If the steps whose numbers have been placed in bold type in Table 2.3 are taken as the starting guess and steady states are introduced for the intermediaries O, OH, HO_2, CH_3, CH_2O, and CHO, then a four-step

Table 2.4. The four-step mechanism for the methane-air flame

Fuel Consumption			
$CH_4 + H$	$\rightarrow$	$CH_3 + H_2$	
$CH_3 + O$	$\rightarrow$	$CH_2O + H$	
$CH_2O + H$	$\rightarrow$	$CHO + H_2$	
$CHO + M$	$\rightarrow$	$CO + H + M$	
$H + OH$	$\rightarrow$	$O + H_2$	
$H + H_2O$	$\rightarrow$	$OH + H_2$	
$CH_4 + 2H + H_2O$	$\rightarrow$	$CO + 4H_2$	
Water-Gas Shift			
$CO + OH$	$\rightleftharpoons$	$CO_2 + H$	
$H + H_2O$	$\rightleftharpoons$	$OH + H_2$	
$CO + H_2O$	$\rightleftharpoons$	$CO_2 + H_2$	
Recombination			
$O_2 + H + M$	$\rightarrow$	$HO_2 + M$	
$OH + HO_2$	$\rightarrow$	$H_2O + O_2$	
$H + H_2O$	$\rightarrow$	$OH + H_2$	
$2H + M$	$\rightarrow$	$H_2 + M$	
Oxygen Consumption and Radical Production			
$O_2 + H$	$\rightleftharpoons$	$OH + O$	
$O + H_2$	$\rightleftharpoons$	$OH + H$	
$OH + H_2$	$\rightleftharpoons$	$H_2O + H$	
$OH + H_2$	$\rightleftharpoons$	$H_2O + H$	
$O_2 + 3H_2$	$\rightleftharpoons$	$2H_2O + 2H$	

mechanism, shown in Table 2.4, is obtained. The extent to which the characteristics of the structure can be discussed on the basis of this four-step mechanism is surprising. Even more surprising is the recent result that many aspects of the flame can be addressed on the basis of a three-step mechanism in which a steady state for H as well is introduced, and actually even on the basis of a two-step mechanism, in which a partial equilibrium for the water-gas reaction is employed in further reducing the three-step mechanism. The two-step mechanism has two stages of burning, first fuel combustion giving the intermediates H_2 and CO,

$$CH_4 + O_2 \rightarrow \left(\frac{2}{1+r}\right)(H_2 + rCO) + \left(\frac{2r}{1+r}\right)H_2O + \left(\frac{1-r}{1+r}\right)CO_2, \quad (2.49)$$

and next the oxidation of the intermediates,

$$\left(\frac{2}{1+r}\right)(H_2 + rCO) + O_2 \rightarrow \left(\frac{2}{1+r}\right)H_2O + \left(\frac{2r}{1+r}\right)CO_2. \quad (2.50)$$

Here r is the ratio of CO to H_2 concentrations at water-gas equilibrium.

The elementary steps whose rates are most important are found to be the branching step 1 (for determining the level of the H-atom concentration), the termination step 5 (for effectively determining the rate of oxidation of H_2), the fuel-attack step 11 (for determining the rate of consumption of CH_4), and the CO-oxidation step 10 (for determining departures from water-gas equilibrium). The resulting structure is illustrated in Fig. 2.8 according to RRA. Following an inert preheat zone, there is a narrow fuel-consumption zone in which the rate of step 11 is important, located where the temperature has a particular value, T_f, determined essentially by equating the rate of the branching step 1 to the geometric mean of the rates of the termination step 5 and the fuel-attack step 11, both of which effectively remove radicals from the system (the latter through the overall result of the fuel chain). A layer of H_2 and CO oxidation, where the rate of step 5 is important, is located downstream from the fuel-consumption zone. For the two-step mechanism, that is all there is to the structure. For the three-step mechanism, there is an additional layer of water-gas nonequilibrium at the upstream end of the layer of H_2 and CO oxidation; under certain conditions of pressure and equivalence ratio, the water-gas nonequilibrium layer may thicken and become larger than the layer of H_2 oxidation, leading to a downstream layer of CO oxidation. For the four-step mechanism, there is an additional layer of radical nonequilibrium in the upstream part of the fuel-consumption layer, as illustrated in the expanded scale at the bottom of Fig. 2.8; although this layer is shown as having a premixed

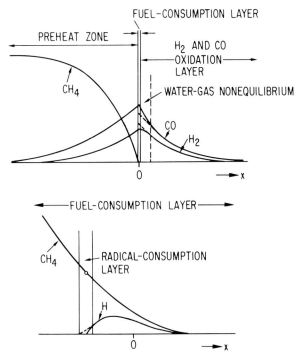

Fig. 2.8. Illustration of the structure of the laminar premixed methane-air flame according to RRA.

character in Fig. 2.8, there are conditions under which it is approximated better as a diffusion flame between CH_4 and H. These various approximations for methane flame structure provide ideas for further improvement in the description.

One contribution of RRA is to provide information on what the overall activation energy E should be if a one-step approximation like (2.2) is arbitrarily imposed. By investigating the dependence of the burning velocity on temperature for stoichiometric methane-air flames, the approximation

$$E = 4R_o T_\infty^2 / (T_\infty - T_f) \qquad (2.51)$$

has been derived, where T_∞ is the adiabatic flame temperature and T_f the fuel-zone temperature identified above. Since T_f depends on rate ratios and T_∞ on thermodynamics, E is found not to be related simply to activation energies of any of the elementary steps. This suggestion broadens ideas about what may contribute to E's measured from temperature dependences of v_o.

Another aspect of the results concerns flame stability. The stability ideas derived from AEA need reevaluation for more complex chemistry.

Some studies have suggested that diffusion of active intermediates can help to stabilize steady, planar flames. Similarly, the constancy of T_f may contribute to stability of the methane flame in the two-step model. Further stability investigations are needed, although recent work has indicated that the AEA results discussed in the Flame Instabilities section of this chapter remain qualitatively correct.

Figure 2.9 is an example of the extent of agreement between theory and experiment for the burning velocity, obtained for hydrocarbon-air flames. This figure, for stoichiometric methane-air flames at different initial temperatures, compares measurements obtained from flame propagation in a closed, spherical chamber with calculations made by RRA using the rates of Table 2.3. The calculations included the first 22 steps in the table, with steady states introduced to yield a three-step mechanism. Since with the structure of Fig. 2.8 all of the fuel chemistry is localized in the fuel-consumption zone, it is relatively easy to add new steps with steady-state approximations for new intermediaries in this approximation. The agreements achieved are not too bad. Agreements this good, or better, generally are also obtained from numerical integrations. This kind of agreement typically extends to fuel-lean flames as well, and the maximum burning velocity is predicted to occur at fuel-rich conditions for chemical-kinetic reasons. The agreement between theory and experiment is poorer for fuel-rich flames, mainly because of the stronger dependence on elementary rates of hydrocarbon chemistry and the poorer accuracies with which these rates are known. It may be concluded that, although recent descriptions of flame structures have

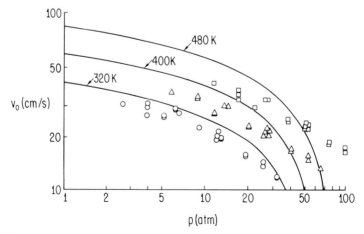

Fig. 2.9. The dependence of the laminar burning velocity on pressure for various initial temperatures for the stoichiometric premixed methane-air flame according to RRA (curves) and experiment (points).

been much more successful than those available ten years ago, there are many aspects in which further improvement is required.

Studies of methane flame structures by RRA include those of Peters and Williams (1987), Seshadri and Peters (1990), and Smooke (1991), the last of which provides a tutorial presentation with entries to the literature. Reduced kinetic mechanisms for other flames as well are given by Peters and Rogg (1993).

Aspects of Deflagrations in Need of Further Investigation

From the preceding discussions it can be seen that studies of pre-mixed-flame structure, propagation, and stability are active areas of research. Many specific uncertainties demanding further investigation have been identified. Table 2.5 is a list of some of the topics currently calling for study by both theoretical and experimental methods. Although there is a very firm foundation, we have much more to learn about deflagrations. Further comments on the summary tables of each chapter, such as Table 2.5, are given in Chapter 6.

Table 2.5. Deflagration problems that need further attention

---

1. Deflagration structure with model and real kinetics
2. Effects of curvature and strain on structure with model and real kinetics
3. Stability of planar, curved, and stretched flames with model and real kinetics
4. Influences of hydrodynamic instability with non-simple kinetics and heat loss
5. Extinction phenomena with real kinetics
6. Nonlinear structures and extinctions at cusps
7. Propagation in gradients of enthalpy or mixture ratio

Bibliography

Barrère, M. and Williams, F.A. 1969. *Twelfth symposium (international) on combustion.* The Combustion Institute, Pittsburgh, pp. 169–81.
Bush, W.B. and Fendell, F.E. 1970. *Combust. Sci. Technol.* **1**, 421.
Chelliah, H.K. and Williams, F.A. 1987. *Combust. Sci. Technol.* **51**, 129.
Clavin, P. 1985. *Prog. Energy Combust. Sci.* **11**, 1.
Darrieus, G. Propagation d'un front de flamme. Paper presented at two conferences in France, La Technique Moderne (1938) and Congrès de Mécanique Appliquée, Paris (1945).
Fristrom, R.M. and Westenberg, A.A. 1965. *Flame structure.* McGraw-Hill, New York.

Glassman, I. 1987. *Combustion*, 2nd ed. Academic Press, New York.

Joulin, G. and Clavin, P. 1979. *Combust. Flame* **35**, 139.

Landau, L.D. 1944. *Zhur. Eksp. Teor. Fiz.* **14**, 240.

Lewis, B. and von Elbe, G. 1987. *Combustion flames and explosions of gases*, 3rd ed. Academic Press, New York.

Markstein, G.H. 1964. *Non-steady flame propagation*. Macmillan, New York.

Mikhel'son, V.A. On the normal ignition rate of fulminating gas mixtures. Ph.D. Thesis, Univ. Moscow (1889); see *Collected works*, vol. 1. Novyi Agronom Press, Moscow (1930).

Peters, N. and Rogg, B., eds. 1993. *Reduced kinetic mechanisms for applications in combustion systems*. Springer, New York.

Peters, N. and Williams, F.A. 1987. *Combust. Flame* **68**, 185.

Rogg, B. 1985. *Combust. Sci. Technol.* **45**, 317.

Rogg, B., Liñán, A. and Williams, F.A. 1986. *Combust. Flame* **65**, 79.

Rogg, B. 1986. *Combust. Flame* **65**, 113.

Seshadri, K. and Peters, N. 1990. *Combust. Flame* **81**, 96.

Smooke, M.D., ed. 1991. *Reduced kinetic mechanisms and asymptotic approximations for methane-air flames*. Springer, New York.

von Kármán, Th. and Millán, G. 1953. *Anniversary volume on applied mechanics dedicated to C.B. Biezeno*. N.V. de Techniche Uitgeverji H. Stam., Haarlem, Holland, pp. 55–69.

von Kármán, Th. and Penner, S.S. 1954. *Selected combustion problems, fundamentals and aeronautical applications*. AGARD, London: Butterworths Scientific Publications, pp. 5–41.

Warnatz, J. 1984. In *Combustion chemistry*, W.C. Gardner, Jr., ed. Springer-Verlag, New York, pp. 179–360.

Williams, F.A. 1985. *Combustion theory*, 2nd ed. Addison-Wesley Publishing Company, Menlo Park, California.

Zel'dovich, Y.B. and Frank-Kamenetskii, D.A. 1938. *Zhur. Fiz. Khim.* **12**, 100.

Zel'dovich, Y.B., Barenblatt, G.I., Librovich, V.B. and Makhviladze, G.M. 1980. *The mathematical theory of combustion and explosion*. Nauka, Moscow; English translation, 1985, Consultants Bureau, A Division of Plenum Publishing Corporation, New York.

3

DIFFUSION FLAMES

Unlike premixed flames, diffusion flames do not have a burning velocity. Interest therefore lies in their internal structure, in their rates of energy release, in rates of transport of fuel and oxidizer into the flame (burning rates), and in necessary conditions for their existence and extinction. For many purposes, diffusion-flame investigations become investigations of transport phenomena. Transport rates are of critical importance in the burning of fuel droplets, in the burning of liquid pools, and in the burning of solid slabs of fuel, for example; two distinct phases (states of matter) must be considered in these processes. There are situations in which diffusion flames spread along surfaces of solid or liquid fuels, or through nonpremixed gases; these flame-spread processes require special consideration. Before considering diffusion-flame structure and extinction, it is of interest to discuss areas of application of diffusion-flame investigations.

Applications of Diffusion Flames

Some of the applications of diffusion flames are listed in Table 3.1. Candles, matches, and wood fires are familiar examples of diffusion flames; heat transfer from the flame to the solid fuel liberates fuel gases that flow and diffuse to the reaction zone where they meet

Table 3.1. Applications of diffusion flames

1. Some stoves and burners
2. Diesel engines
3. Oil-fired boilers
4. Wood fires
5. Coal burners
6. Fire spread rate prediction
7. Fire suppression strategies

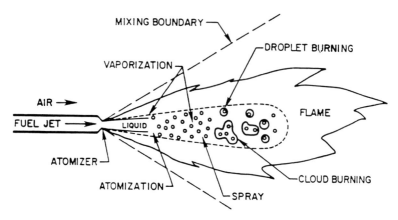

Fig. 3.1. Schematic illustration of spray combustion for a liquid fuel injected into a gaseous oxidizer.

oxygen and burn. The rate at which the fuel is consumed—the burning rate—depends on the rate of heat transfer. The rate at which coal burns in a coal furnace similarly depends on the rate of heat transfer. The small blue flames in gas stoves used, for example, in home cooking, often have a dual structure consisting of fuel-rich premixed flames surrounded by diffusion flames; their lengths depend on the gas flow rate, and knowledge of both premixed and diffusion flames is helpful in the design of burners for stoves. Gas-fired furnaces and boilers involve premixed flames or diffusion flames, depending on whether the fuel is premixed with air. While spark-ignition engines, such as the Otto engine, involve propagation of premixed flames, the Diesel engine employs diffusion flames that ignite and burn around droplets or sprays. Stratified-charge engines are designed for premixed combustion in a varying mixture ratio (the last entry in Table 2.5), although especially with direct injection they also may contain regions of diffusion-flame combustion. Oil-fired boilers involve diffusion-flame combustion of droplets or sprays. In many liquid-propellant rocket engines diffusion-flame combustion occurs around individual fuel or oxidizer droplets, and there are combustion processes in solid-propellant rockets that involve diffusion flames, such as the burning of metal particles in oxidizing gas. Supersonic combustion, considered for hypersonic propulsion, typically involves diffusion-flame combustion of hydrogen injected into air. Knowledge of diffusion flames is important for predicting rates of fire spread and also in strategies for fire suppression and in the selection and use of fire extinguishers since to extinguish a fire it is necessary to cause extinction of its gas-phase diffusion flame.

Table 3.2. Spray-combustion processes

---

1. Atomization
2. Vaporization
3. Heat transfer
4. Droplet-air mixing
5. Vapor-air mixing
6. Ignition
7. Turbulent diffusion flames
8. Premixed combustion
9. Production of NO_x and of other pollutants
10. Extinction

Processes involved in a representative example of diffusion-flame combustion can be addressed by considering the spray combustion of a liquid-fuel jet issuing into an oxidizing gas. This process is illustrated in Fig. 3.1. As summarized in Table 3.2, elements of the process may include atomization of the liquid jet (which may occur by many different mechanisms, depending on the application), vaporization of the liquid fuel, heat transfer from the flames to the liquid to cause vaporization (and also to the surroundings), mixing of the droplets with the oxidizing gas, mixing of the fuel vapor with the oxidizer, ignition in the nonuniform gas, diffusion flames around individual droplets or around clouds of droplets (cloud burning), possibly some regions of premixed-flame propagation, production of air pollutants such as oxides of nitrogen by chemical-kinetic processes in the hot flames, and possibly the extinction of burning around fuel droplets as they reach cold gas, leaving unburnt fuel that can represent additional pollutants and detract from the overall efficiency of combustion. Quantitative prediction of histories of all of these processes of spray combustion would be quite difficult. Progress is made by addressing different aspects separately. One aspect is just the post-atomization, post-mixing stage of flow and combustion of a two-phase, droplet-gas mixture.

The time-dependent conservation equations, (1.1) through (1.4), in principle can be applied to describe the flow and combustion of the droplet-gas mixture. In practice this is prohibitively difficult because it necessitates determining whether liquid or gas is present at every point and employing a different equation of state, for example, depending on which phase is present. Generally there are so many droplets that a description at this level cannot be achieved, and the droplets must be treated statistically. For this purpose, a spray density function $f_j(r, t, x, v, \cdots)$ may be employed, with $f_j dr dx dv \ldots$ defined as the number of droplets of chemical type j having radius in the range dr about r and velocity in the range dv about v, located in the volume element dx about x at time t. Depending on the problem, it may be desirable to incorporate additional variables into f_j in the description of the droplets, such as parameters measuring departures from spherical symmetry, or droplet temperature. A conservation equation for f_j, the "spray equation," can be written from phenomenological reasoning, in analogy with the kinetic theory of gases, as

$$\frac{\partial f_j}{\partial t} + \frac{\partial}{\partial r}(R_j f_j) + \frac{\partial}{\partial x} \cdot (\boldsymbol{v} f_j) + \frac{\partial}{\partial \boldsymbol{v}} \cdot (\boldsymbol{a}_j F_j) = Q_j + P_j , \qquad (3.1)$$

where R_j is the rate of increase of radius of the droplet, $\boldsymbol{a}_j$ is its acceleration, Q_j represents the number produced per unit time (per unit volume, per unit radius range, per unit velocity range), for example from injection or atomization, and P_j is this same production rate resulting from droplet collisions. This particular description would be fairly general; simplifications occur in many regions, for example away from droplet sources and where the droplets are far enough apart that their collisions can be neglected, giving zero on the right-hand side. The main point is that some sort of statistical description of the droplets generally must be introduced.

Beyond this complication, to describe the combustion process conservation equations are then needed for each phase separately. In these problems the droplets are dispersed throughout the gas, so they form a "dispersed phase" in the "continuous phase" of the gas. Many complexities arise in trying to define local average properties of the continuous and dispersed phases for stating conservation laws. For example, because of the presence of the droplets, the local average gas density per unit volume of the gas is greater than the local average gas density per unit volume of space; mass conservation for the latter quantity contains a source term from droplet evaporation, which involves an integral of the spray density function. Integrodifferential conservation equations therefore are obtained. For dilute sprays—those in which the volume occupied by droplets is small compared with that occupied by gas—it is convenient to use these gas conservation equations and results for R_j, $\boldsymbol{a}_j$, etc., derived from analyses of evaporation (or burning) and acceleration of single droplets in nonuniform gas flows to obtain a closed set of equations. There are spray-combustion problems for which this set has been integrated to give descriptions of the process.

The complexities of multiphase flows are thus seen to pose challenges in deriving suitable methods of description. For many diffusion-flame situations it is important to have these descriptions for use in applications. Some notable successes have been achieved, but in view of the magnitude of the difficulties, relatively little effort has been devoted to multiphase questions, and much remains to be done. Problems of this kind are not addressed further here. Instead, attention will be focused on more elementary problems, such as finding the burning rate of an individual droplet in an oxidizing gas—results needed as inputs to descriptions of multiphase combustion. Wallis (1969), Drew (1983), Faeth (1983), and Williams (1985) provide entries into the literature on multiphase flow.

Structures of Laminar Diffusion Flames

The same kinds of experimental diagnostic techniques that have been described for measuring premixed-flame structures also can be applied to the investigation of laminar diffusion-flame structure. A further complication for diffusion flames is that the flow is never one-dimensional. Therefore, instrumentation for measuring gas velocities is relatively more important for diffusion flames (although also important for nonplanar premixed flames). The older Pitot-tube and hot-wire methods can be used to some extent but encounter severe difficulties because of the large changes in density and temperature. The advent of laser-Doppler velocimetry, in which the flow is seeded with fine particles, and the velocity is obtained from the Doppler effect in laser scattering from the particles, has greatly increased experimental capabilities for velocity measurements in flames. Seeding techniques have progressed considerably in recent years to make certain that the seed particles are small enough to follow the gas velocity and dilute enough to interact negligibly and to exert a negligible influence on the flame. Even without velocity measurement, seeding has proven useful in combustion diagnostics, for example in showing streamlines and in marking isotherms such as flame-sheet locations in premixed flames. In the latter technique, particles or droplets that vaporize at a particular temperature are employed, so that illumination by a laser sheet shows a spatial contour of the isotherm (laser tomography). These advances in measurement capabilities have greatly improved our experimental knowledge of flame structures.

To investigate diffusion-flame structures it is desirable to establish a simple configuration that can be probed in detail. The counterflow diffusion flame represents such a configuration. Opposing jets of fuel and oxidizer are directed toward each other, and after ignition (by a spark or a match) a planar diffusion flame is established normal to the axis of the jets. The jet flows are tailored to exhibit uniform exit velocity profiles, and the velocities are large enough to make buoyancy effects negligible (Froude number large enough), but small enough to prevent flow instabilities and turbulence from developing (Reynolds numbers not too large). This approach also is readily applied to liquid or solid fuels; the gaseous oxidizing jet is directed downward onto the flat surface of a liquid-fuel pool or a solid-fuel slab, and the flat flame is established above the fuel surface, with the fuel feed adjusted to maintain a fixed location of the fuel surface in the laboratory. Because the flame and the isotherms are steady and horizontal in the laboratory, the older thermocouple and gas-sampling techniques may be employed to measure temperature and composition profiles, with the intrusive probes inserted horizontally along isotherms; diffusion flames are thicker than premixed flames, typically a few millimeters thick in these experiments, and therefore the difficulties in spatial resolution are less severe, so that accurate measurements can be made even at normal atmospheric pressures and above.

For liquid or solid fuels, an alternative experiment is to study the burning of spherical droplets or solid fuel particles such as metal spheres. These experiments are not steady because the fuel surface recedes with time as it burns, and although burning times and histories can be observed by motion-picture photography, seldom can temperature or composition profiles be obtained with reasonable accuracy. Also, unless the sphere radius is small (typically less than 0.1 mm), buoyancy causes the flame to be distinctly nonspherical. Yet, because of practical interest in the results, many experiments of this kind have been performed. The possible future availability of a combustion laboratory in a space vehicle may enable measurements of this kind to be accomplished more accurately by reducing the complications of buoyancy.

A significant difference between premixed flames and diffusion flames is that in the former the fuel-air mixture ratio is everywhere constant, while in the latter it varies. Thus while the progress of combustion in premixed flames can be described by a variable measuring the extent of completion of the reaction, in diffusion flames an additional variable measuring the mixture ratio is needed. A convenient selection for the latter is the mixture fraction Z, defined to be zero in the oxidizer stream and unity in the fuel stream, and locally to be the fraction of the total amount of material present that came originally from the fuel stream. The mixture-fraction field $Z(x, t)$ is one of the most important descriptors of a diffusion flame. In the counterflow flame, Z is a function only of the distance along the axis. In other configurations, the direction of its gradient (∇Z) may be called the local direction normal to the flame. A coordinate transformation from (x, t) to (y, Z, t) often is useful in describing diffusion-flame structure. Here the coordinate normal to the flame is replaced by Z, and y represents the two remaining spatial coordinates, those tangent to the flame. Profiles of temperature and of concentrations in diffusion flames may be shown as functions of Z instead of as functions of the distance normal to the flame. An advantage of the transformation is that in the Z coordinate the profiles generally are much less strongly dependent on the flow configuration than they are in the original spatial coordinate. To the extent that the profiles depend only on Z, results of investigations of counterflow diffusion flames may be applied to other configurations as well after $Z(x, t)$ is found for those configurations.

For diffusion flames the fast-chemistry limit, which corresponds to large values of the first similarity group of Damköhler (1936), $D_k = \infty$ for all k in (1.4), is a useful limit to consider because it can possess uniform validity (that is, it can hold throughout space and time). For premixed flames this limit cannot be uniform since it would imply that the reactants would already have reacted wherever they may be. With fast chemistry, chemical equilibrium is maintained everywhere, and chemical compositions then vary only because of variations in pressure, temperature, and mixture fraction; values of reaction-rate constants no longer affect the flow. Equilibrium constants are known more accurately than reaction-rate constants. Techniques for calculating equilibrium compositions are well developed and available in computer routines. Therefore the greatest sources of uncertainty in computation of fast-chemistry diffusion-flame structure lie in the transport coefficients. The problem of describing the diffusion-flame structure becomes a problem in heat and mass transfer for a multicomponent fluid with complicated thermodynamic properties.

This transport problem is simplified significantly if pressure-gradient, body-force, and thermal diffusion are negligible and if the binary diffusion coefficients of all pairs of species are approximately equal. Fick's law (1.8) then applies under these equidiffusional conditions, and from (1.4) it can be shown that the concentrations of all species are uniquely related to Z through algebraic equations. If this condition is not satisfied, then differential diffusion can cause definitions of Z based on different elements (N, O, C, H, etc.) to give different values, sometimes negative or greater than unity. In gases, departures from equidiffusional behavior usually are small enough to be investigated by perturbation methods, with (1.8) used in a first approximation. If M and ω are small and B is large in (1.3), and if the thermal diffusivity equals the binary diffusion coefficient (Lewis number of unity), then the temperature as well is found to depend only on Z, and diffusion-flame profiles with Z as the coordinate are the same in all configurations. An approximation to these profiles is obtained by employing a one-step

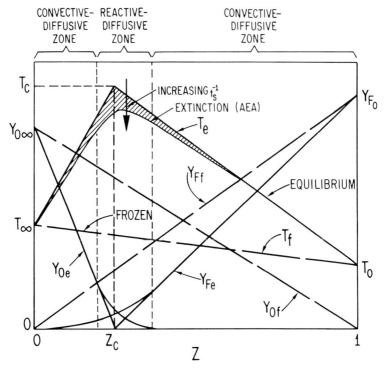

Fig. 3.2. Illustration of the structure of a diffusion flamelet in mixture-fraction space.

representation of the combustion process. The result is shown in Fig. 3.2 as the equilibrium lines.

   In Fig. 3.2 Z_c is the stoichiometric mixture fraction; the mixture is fuel-lean for $Z < Z_c$ and rich for $Z > Z_c$. With the one-step rate function (2.2), for infinite Damköhler number either Y_F or Y_O must be zero, and this is possible only with $Y_F = 0$ for $Z < Z_c$ and $Y_O = 0$ for $Z > Z_c$. There is a singularity at $Z = Z_c$ where fuel and oxidizer are consumed at an infinite rate per unit volume, and all of the chemical heat of reaction is released. Thus there are two separate zones, one $(Z < Z_c)$ in which oxygen diffuses into the flame and the other $(Z > Z_c)$ in which fuel diffuses into the flame (hence the name "diffusion flame"). Since the reaction rate vanishes in these zones, both are convective-diffusive zones when the flows are steady.

   The reaction rate must be important in the flame itself, but to see the structure the scale must be expanded about $Z = Z_c$. When such a stretching of coordinates is made in the conservation equations, the most highly differentiated terms become large compared with lower derivatives. Therefore the accumulation and convection effects are small

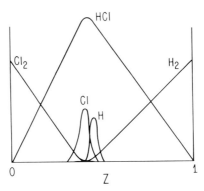

Fig. 3.3. Illustration of the structures of hydrogen-chlorine diffusion flames.

here compared with the diffusion effects, and a diffusive-reactive balance is established, in a first approximation. The flame zone thus is a reactive-diffusive zone. From (1.4), the first approximation to the differential equation describing its structure is

$$- \rho D_{12} |\nabla Z|^2 \, d^2 Y_i / dZ^2 = w_i \,, \tag{3.2}$$

where w_i is the mass rate of production per unit volume of species i in the flame, and D_{12} is the common binary diffusion coefficient. Boundary conditions for (3.2) are that the slopes dY_i/dZ must approach those shown in Fig. 3.2 as $Z - Z_c$ approaches $\pm \infty$, according to the method of matched asymptotic expansions. The situation clearly is analogous to that discussed in the previous chapter for premixed-flame structure.

For a one-step approximation in the fast-chemistry limit, scaling of variables is found to produce the inner problem

$$d^2 y / d\zeta^2 = (y + \zeta)^m \, (y - \zeta)^n \,, \quad dy/d\zeta \to \pm 1 \text{ as } \zeta \to \pm \infty, \tag{3.3}$$

where $\zeta \sim Z - Z_c$ is the stretched mixture-fraction variable, $y + \zeta$ the scaled fuel concentration, $y - \zeta$ the scaled oxidizer concentration, and m and n the reaction orders appearing in (2.2). Numerical solution of (3.3) is readily completed to show the continuous variations of fuel and oxidizer concentrations through the reaction zone. The question arises as to effects of detailed chemistry of real flames on this simple structure. In principle, with more complicated chemistry the single thin reaction zone at $Z = Z_c$, shown in Fig. 3.2, need not occur. As a hypothetical example consider the hydrogen-chlorine flame, whose chain-reaction mechanism is similar to that given in the previous chapter for the hydrogen-bromine flame. Although it is conceivable that the two main chain-carrying steps, $H + Cl_2 \to HCl + Cl$ and $Cl + H_2 \to HCl + H$, may occur at different locations, with radical diffusion present between these two separated reaction zones, study shows that this in fact does not occur and that there

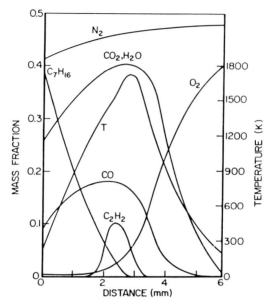

Fig. 3.4. Schematic illustration of the structure of a laminar diffusion flame of C_7H_{16} in an O_2-N_2 mixture.

is a single reaction zone instead, containing a region of chlorine dissociation on its oxidizer side, as illustrated in Fig. 3.3. However, detailed
measurements of structures of hydrocarbon-air flames have provided evidence for departures from structures predicted by one-step chemistry,
and numerical computations (Bui-Pham and Seshadri, 1991, unpub-
lished) of structures of diffusion flames between CH_4 and NO_2, overall
$CH_4 + 2NO_2 \rightarrow CO_2 + 2H_2O + N_2$, have identified a four-step reduced
mechanism and have shown widely separated zones of fuel consumption and of oxidizer consumption, like those first guessed for H_2 and
Cl_2.

Figure 3.4 is an illustration of flame structures measured for hydrocarbon-air flames. In this figure the fuel is heptane, and profiles of temperature and concentrations are shown as functions of the vertical distance normal to the liquid-fuel surface for a counterflow experiment.
The fuel vapor and oxygen are seen to diffuse into a thin reaction zone,
located about 2.5 mm above the liquid surface. The temperature and
concentrations of the final product tend to peak at this location, but
intermediate products CO (and H_2, not shown) are present in appreciable concentrations and are not completely consumed, but instead penetrate to some extent into regions too cold to oxidize them. A telltale
observation is that none of the original fuel leaks to the air side, but
oxygen concentrations on the order of 1% survive all the way to the fuel

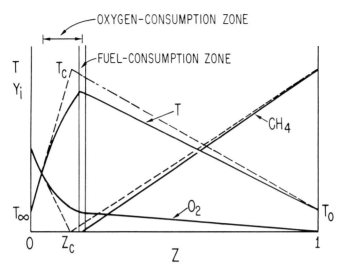

Fig. 3.5. Illustration of the structure of the methane-air diffusion flame according to RRA.

surface. The preferential penetration of O_2 is inconsistent with one-step chemistry. However, it is consistent with a two-step mechanism of the type described in the previous chapter for the methane flame. This is illustrated in Fig. 3.5 for methane, which experimentally exhibits about the same amount of O_2 leakage as heptane.

In Fig. 3.5, the dashed lines represent the equilibrium structure as shown in Fig. 3.2, with a one-step approximation for the chemistry. The two-step approximation given in Chapter 2 produces a structure that approaches that of the one-step approximation for infinite Damköhler numbers but that differs, as illustrated by the solid lines in Fig. 3.5, for finite Damköhler numbers. The temperature peaks in a narrow fuel-consumption zone around $Z = Z_f > Z_c$ where the fuel is oxidized partially to H_2O and CO_2 and partially to H_2 and CO. The reaction ceases on the rich side of this zone because of depletion of radicals, not because of complete consumption of oxygen, and therefore some O_2 leaks through this zone. On the lean side of the fuel-consumption zone, extending to $Z < Z_c$, is a zone of primary oxygen consumption and of combustion of the intermediates, H_2 and CO, which are depleted nearly completely before reaching the lean convective-diffusive zone. The structure resembles that described in the previous chapter for the premixed flame, with the reactant side corresponding to $Z = 1$ and the product side to $Z = 0$, although the temperature profiles, of course, are quite different. The peak temperature in Fig. 3.5 is seen to be less than the adiabatic flame temperature, in agreement with Fig. 3.4, where the maximum value is about 300 K below the adiabatic flame temperature; this difference is attributable mainly to the finite-rate chemistry and is reflected in diffusion of O_2, CO, and H_2 into the convection-diffusive zone at the fuel side. Current research is devoted to finding how well the diffusion-flame structure may be described by this simplified mechanism or by extensions or modifications thereof. Representative recent contributions are those of Seshadri and Peters (1990) and of Chelliah and Williams (1990). This research is aided by numerical integrations of the full set of conservation equations for the counterflow diffusion-flame structure with complete chemistry and with various reduced mechanisms; although these integrations are more difficult to perform than those for the corresponding premixed flame, many have been accomplished.

Other aspects of the flame structure shown in Fig. 3.4 concern the slight indentation of the temperature profile on the fuel side of the peak and the acetylene concentrations there well in excess of equilibrium. The C_2H_2 profile shown is roughly representative of profiles (not shown) of many nonequilibrium fuel species present on the fuel side of stoichiometry (Saito et al., 1986). These species arise from fuel pyrolysis (breakdown and further reactions of the fuel molecules under the influence of heat), and the pyrolysis may absorb heat, producing the temperature indentation often observed and sometimes called the "pyrolysis dip." These fuel pyrolysis reactions are quite complex and also specific to the particular fuel, although some species, such as C_2H_2, apparently always are formed to some extent in hydrocarbon pyrolysis. After fuel breakdown, the reactions also proceed to build molecules larger than the original fuel molecule. When residence times on the fuel side of the flame are large enough (large Damköhler numbers), this fuel chemistry eventually leads to soot production (Glassman, 1987; Howard, 1991).

With short residence times, hydrocarbon flames are predominantly blue because of (nonequilibrium) chemiluminescent radiation from electronically excited CO, CH, C_2, etc., formed by chemical reactions in the main high-temperature reaction zone, but the yellow radiation from soot particles on the fuel side of the flame can obscure the blue completely at large residence times. The soot particles, once formed, can grow rapidly by chemically annexing smaller molecules, such as C_2H_2, and they, too, may be oxidized by radicals or O_2 in the hottest part of the flame. The hydrocarbon fuel chemistry starts at roughly 700 K and typically extends throughout most of the fuel zone, although often as a perturbation, so that the zone remains mainly convective-diffusive. So wide a range of chemical times is involved that, in practice, the limit $D_k \to \infty$ for every step is unachievable; increasing residence times too much in hydrocarbon diffusion flames usually results in heavy sooting, in soot agglomeration, and in the consequent introduction of chemical processes with even longer chemical times. Rate parameters for the large-molecule chemistry are poorly known. Even if they were known, inclusion of all of them would exceed computer capacities for numerical integrations. Therefore the large-molecule chemistry is not part of the current computational arsenal. Obtaining simplified kinetic descriptions of rates of fuel pyrolysis and of soot production is a further objective of current research on hydrocarbon diffusion flames (Seshadri et al., 1991).

Adding chemicals to the flow, such as Br_2 or CF_3Br (a useful fire suppressant for chemical fire extinguishers that is now being eliminated because of influences of fluorocarbons on the earth's ozone layer), modifies the diffusion-flame structure. The modifications can be measured in the counterflow burner. Many of these chemical suppressants, in addition to slowing the chemistry in the primary reaction zone by reducing radical concentrations there (for example, through $H + Br_2 \rightarrow HBr + Br$, with Br less active than H), modify the chemistry in the fuel zone, enhancing soot production. Counterflow hydrocarbon flames without additives generally lose their flatness through buoyancy before residence times can be increased enough to produce heavy sooting. The coflow flame, in which the fuel flows vertically from a tube into quiescent air or into a coflowing oxidizer stream, provides longer residence times at readily achievable tube diameters (~ 1 cm) and exit velocities (~ 50 cm/s) for laminar flow in the laboratory. It is this flame whose shape was first derived through theoretical analysis in the one-step reaction-sheet approximation by Burke and Schumann (1928). Experimentally, as the exit velocity is increased, the residence time increases (in fact not according to the Burke-Schumann analysis but instead because of buoyancy, which they did not include; see, for example, Williams, 1985), and a "soot point" is reached at which soot begins to penetrate the top of the flame and escapes, unburnt. Although influenced significantly by the flow field as well as by chemical kinetics, the soot point (or the "soot height," the height of the yellow flame at the soot point) is a good indicator of relative sooting tendencies of different hydrocarbon fuels. Another is the (much lower) exit velocity just above which yellow radiation from soot first appears.

Diffusion-Flame Extinction

An alternative to the expansion for large Damköhler numbers is an (AEA) expansion for large Zel'dovich numbers in (3.2). This is easiest for the one-step approximation of (2.2) and has been given for the counterflow flame by Liñán (1974). In Fig. 3.2, the dashed lines along which properties are identified by the subscript f (frozen) represent the nonreactive mixing in which the fuel and oxidizer interdiffuse without combustion. An appropriate Zel'dovich number is

$$\beta = E(T_c - T_{fc})/[2Z_c(1 - Z_c)R_oT_c^2], \tag{3.4}$$

where T_{fc} is the frozen temperature at $Z = Z_c$. For $m = n = 1$, with

$$\zeta \equiv \beta(Z - Z_c), \quad y \equiv (T_c - T)E/(R_oT_c^2) - \gamma\zeta, \tag{3.5}$$

in which

$$\gamma \equiv 2Z_c - 1 - 2Z_c(1 - Z_c)(T_{Fo} - T_{Oo})/(T_c - T_{fc}), \tag{3.6}$$

the inner equation obtained from (3.2) is, in the first approximation,

$$d^2y/d\zeta^2 = \delta(y^2 - \zeta^2)e^{-(y+\gamma\zeta)}, \tag{3.7}$$

with the matching conditions $dy/d\zeta \to \pm 1$ as $\zeta \to \pm\infty$. Here the reduced Damköhler number is

$$\delta = (Y_{Oo}A_Fe^{-E/R_oT_c})\big/(2\rho D_{12}|\nabla Z|_c^2 Z_c\beta^3), \tag{3.8}$$

in which A_F is the rate constant of (2.2) for the mass rate of consumption of fuel, and the subscript c on the gradient means that it is evaluated at $Z = Z_c$. The solution to (3.7) depends on the two parameters, δ and γ, the last of which is a measure of the ratio of the temperature gradient on the fuel side to that on the oxidizer side at equilibrium. For a representative value of γ, the dependence of the maximum temperature T_m on δ, obtained by numerical integration of (3.7), is illustrated schematically in Fig. 3.6.

For large values of δ, rescaling of coordinates can be made in (3.7) to produce (3.3) (with $m = n = 1$), and the same flame structure for large Damköhler numbers therefore is obtained. However, in a range of δ of order unity, the variation of δ with T_m is nonmonotonic in Fig. 3.6; there are three values of T_m for a given value of δ, each corresponding to a different set of temperature and composition profiles. At still smaller values of δ, again there is only one solution, that corresponding to nonreactive mixing in Fig. 3.2. In the "S-shaped curve" of Fig. 3.6, the upper branch corresponds to the presence of the flame and the lower branch to the absence thereof; the middle branch represents an unstable configuration that does not occur in real flames. A gradual increase of δ from small values is predicted to produce a sudden ignition at δ_I,

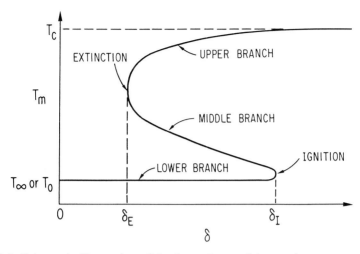

Fig. 3.6. Schematic illustration of the dependence of the maximum temperature on the reduced Damköhler number for diffusion flames.

and a gradual decrease from large values, a sudden extinction at δ_E. For ignition the expansion leading to (3.7) cannot be justified, and an alternative analysis is needed, but for extinction the expansion often leads to reasonably accurate predictions. Usually the magnitude $|\gamma|$ is small, and from an expansion for small $|\gamma|$ it is found that at extinction

$$\delta_E = 1 - |\gamma|. \tag{3.9}$$

Of course, the analysis does not describe the dynamics of extinction, which would require consideration of time-dependent equations. Also, instability may occur on the curve slightly away from δ_E. If one of the boundary temperatures exceeds T_c, then a monotonic variation of T_m with δ is found instead of Fig. 3.6, δ_E does not exist, and extinction is gradual, not abrupt. Finally, there are two other possible scalings (besides the ignition and diffusion-flame scalings), one of which, applicable near extinction (when Z_c or $1 - Z_c$ is small), corresponds to substantial leakage of one reactant or the other through the reaction zone; the analysis of this "premixed-flame regime" is the same as that needed for an AEA description of extinction of premixed flames through heat loss to the burnt gas.

In the counterflow experiment, increasing the exit velocity U_e from the duct increases the strain rate of the flow, which is proportional to U_e/ℓ, where ℓ is the duct spacing. In the flow field this produces a corresponding increase of $D_{12}|\nabla Z|_c^2$ in (3.8), thereby decreasing the

DIFFUSION FLAMES

Damköhler parameter. An effective strain time can be defined as

$$t_s = \left(D_{12} \, |\boldsymbol{\nabla} Z|_c^2 \right)^{-1} , \qquad (3.10)$$

so that increasing the strain rate decreases the Damköhler number by decreasing t_s. Figure 3.2 illustrates the predicted effect of this decrease on the temperature profile $T(Z)$. Relatively small departures from the equilibrium profile occur before extinction; after extinction essentially only the frozen profile is observed, so that a large region of the figure is excluded. This abruptness in behavior is a consequence of β being large (typically $10 \lesssim \beta < 100$); a fractional temperature reduction on the order of β^{-1} extinguishes the flame because of the strong dependence of the rate on T. Experimentally it is observed that when U_e is increased past a critical value, the flat flame suddenly is extinguished, in agreement with prediction. The theory can be used along with measurements of U_e/ℓ at extinction to calculate overall rates of heat release. Dilution of the oxidizer with an inert gas like N_2 decreases the adiabatic flame temperature T_c. From measurements of U_e/ℓ at extinction as a function of dilution, an Arrhenius plot can be

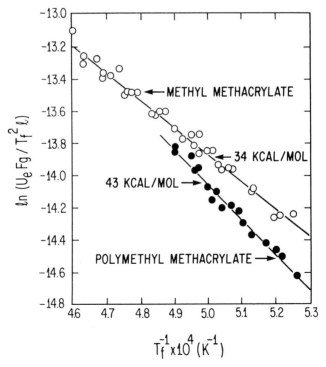

Fig. 3.7. Arrhenius plot of data for flame extinction of liquid methyl methacrylate (open symbols) and of solid polymethyl methacrylate (solid symbols).

constructed for extracting both of the rate parameters E and A_F (Williams, 1981). Figure 3.7 is an example of such a plot; the nondimensional quantity Fg is obtained from the theory, and the slope on the semilog graph is $-E/R_o$. This extinction experiment thus provides an alternative to burning-velocity measurements for obtaining overall reaction-rate parameters.

As has been discussed in Chapter 2 for premixed flames, the meaning of experimental results for E and A_F needs to be investigated in terms of the detailed chemistry occurring in the flame. Currently there is intensive research on this question for hydrocarbon-air diffusion flames. Since strong leakage of one reactant is predicted by the AEA analysis in the premixed-flame regime, it might be thought that the O_2 behavior in Fig. 3.4 could be consistent with one-step chemistry. However, on the contrary, since Z_c is small for these flames ($0.03 \lesssim Z_c \lesssim 0.10$, depending on dilution), the one-step theory favors fuel leakage. At least a two-step mechanism, like that of Fig. 3.5, is needed to predict the O_2 leakage. Extended to a three-step mechanism, RRA analysis for the methane-air flame has produced the curve in Fig. 3.8 (Seshadri and Peters, 1988), giving the maximum temperature as a function of the strain time t_s of (3.10). In this figure, the triangles are results of numerical integrations with a four-step mechanism, and the squares are results of measurements. The RRA approach predicts abrupt extinction for chemical-kinetic reasons, without AEA. Although order-of-magnitude agreements for t_s at extinction are reasonable, notable discrepancies and uncertainties remain, and more research is needed.

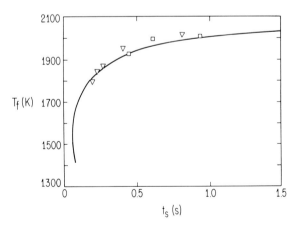

Fig. 3.8. Dependence of the maximum temperature on the effective inverse strain rate for methane-air diffusion flames, according to asymptotic analysis with a three-step mechanism (curve), numerical integration with a four-step mechanism (triangles), and experiment (squares).

Droplet Burning

A spherical fuel droplet in an oxidizing atmosphere burns as illustrated in Fig. 3.9. There is a reaction zone in the gas, where fuel and oxidizer are consumed to release heat, which is partially conducted back to the liquid surface to vaporize the fuel. A representative flow time may be constructed from the droplet radius r_ℓ and a diffusion coefficient D_{12} as r_ℓ^2/D_{12}, which replaces (3.10). Since this time decreases as r_ℓ decreases, when r_ℓ becomes small enough flame extinction is anticipated, for the reasons just explained, if the liquid temperature T_ℓ and the ambient temperature T_∞ are less than T_c. Although such extinctions have been observed under certain conditions (Cho, Choi, and Dryer, 1991), sufficiently detailed measurements of them have not yet been obtained to extract overall rate parameters. In fact, other phenomena often intercede before extinction and prevent it from being measured. One such phenomenon is droplet disruption through internal vaporization. The surface temperature of the liquid is limited by the boiling point of the mixture at the surface, and for multicomponent fuels this increases with decreasing volatility (increasing concentration of the larger molecules). As combustion progresses, the smaller fuel molecules vaporize more readily, and the larger ones accumulate in a boundary layer in the liquid at the surface, with thickness on the order of the liquid-phase diffusion coefficient divided by the linear regression rate of the surface. The surface temperature therefore can increase with time, and since

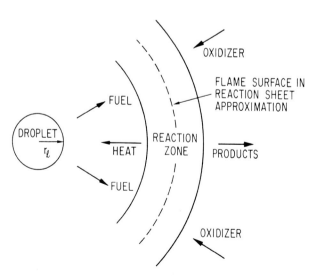

Fig. 3.9. Model of the burning of a fuel droplet in an oxidizing atmosphere.

thermal diffusivities of liquids are large compared with their molecular diffusivities, the internal temperature of the droplet increases and may eventually exceed the homogeneous nucleation temperature of the interior mixture. When this happens, a bubble develops in the liquid and expands rapidly, causing the droplet to shatter. Also, for heavy fuels the liquid temperature may become high enough for fuel pyrolysis to occur in the liquid. For pure fuels with low enough boiling points, these processes are absent, and observation of extinction may be anticipated.

However, there is a complicating factor in droplet burning that still may prevent gas-phase extinction from occurring before completion of vaporization. Although the time estimate r_ℓ^2/D_{12} is appropriate for quasisteady combustion, because the process is initiated at time $t = 0$ there is an outer transient-diffusive zone at radii greater than $\sqrt{D_{12} t}$, and depending on the stoichiometry, the flame may lie in this outer zone. If so, then the characteristic residence time increases with time until vaporization is completed, and extinction cannot occur until the flame expansion ceases, after vaporization. Experiments on droplet burning with nonspherical convection minimized show a flame radius initially increasing with time and often never decreasing during the observation time. These experiments are performed in falling chambers to reduce buoyancy, which is a major cause of nonspherical convection; after the experiment is released to free fall, the droplet is ignited by a spark and the combustion recorded photographically. Droplets burning above the critical pressure of the liquid, a condition that often is of practical importance, for example in Diesel engines, always experience this transient type of behavior, although an inner fuel radius still may exhibit quasisteady behavior because of the strong decrease in the fuel diffusion coefficient with decreasing temperature.

The burning rate of a droplet, measured by the rate of decrease of droplet radius, is needed in combustor design, for example for use in (3.1). In the quasisteady stage after heatup, this can be estimated by investigating the mixture-fraction field Z. Under the previously indicated assumptions that allow the concentration and temperature fields to be related to Z, from (1.4) the general conservation equation for Z is found to be

$$\partial \left(\rho Z\right)/\partial t + \boldsymbol{\nabla} \cdot \left(\rho \boldsymbol{v} Z\right) = \boldsymbol{\nabla} \cdot \left(\rho D_{12} \boldsymbol{\nabla} Z\right), \qquad (3.11)$$

since the chemical source vanishes by the definition of Z. If the ratio of the gas density to the liquid density is small, then in the first approximation the surface regresses at a velocity small compared with the gas velocity, and in spherical symmetry the quasisteady equation

$$d\left(r^2 \rho v Z\right)/dr = d\left(r^2 \rho D_{12} dZ/dr\right)/dr \qquad (3.12)$$

is obtained, with $Z = Z_\ell$ at $r = r_\ell$ and $Z = 0$ at $r = \infty$.

Mass conservation (1.1) shows that the rate of mass loss by the droplet is

$$\dot{m} = 4\pi r^2 \rho v . \qquad (3.13)$$

Use of this in (3.12) enables the equation to be integrated if ρD_{12} is constant. The further boundary condition of energy conservation at the droplet surface,

$$\dot{m} L = \left(4\pi r^2 \lambda dT/dr\right)_\ell , \qquad (3.14)$$

where L is the energy required per unit mass to vaporize the fuel, serves to determine $\dot{m}/(4\pi r_\ell \rho D_{12})$ in terms of Z_ℓ. Results can be expressed in terms of the transfer number,

$$B \equiv \left[Q_O Y_{O\infty} + c_p\left(T_\infty - T_\ell\right)\right]/L , \qquad (3.15)$$

as

$$\dot{m} = 4\pi r_\ell \rho D_{12} \ln\left(1 + B\right) , \qquad (3.16)$$

and

$$Z = 1 - \left(1 + B\right)^{-r_\ell/r} . \qquad (3.17)$$

Here the gas specific heat c_p has been assumed constant, and Q_O is the energy released in combustion per unit mass of oxidizer consumed. The transfer number B, which appears here, represents the ratio of a thermodynamic driver for mass transfer to a thermodynamic resistance and is also given by

$$B = \left(C + Y_{F\ell}\right)/\left(1 - Y_{F\ell}\right) , \qquad (3.18)$$

where $Y_{F\ell}$ is the fuel mass fraction in the gas at $r = r_\ell$, and

$$C \equiv Z_c/\left(1 - Z_c\right) \qquad (3.19)$$

is the stoichiometric mass ratio of the fuel stream to the oxidizer stream. The quasisteady ratio of the flame radius to the droplet radius in the reaction-sheet approximation is found to be

$$r_c/r_\ell = \ln\left(1 + B\right)/\ln\left(1 + C\right) . \qquad (3.20)$$

The transfer number B of (3.15) and (3.18) arises not only in droplet-burning theory but in fact in all analyses of rates of mass transfer from one phase to another (Kanury, 1975; Rosner, 1986). It is basic to general formulations of the theory of mass transfer. According to (3.17), the mixture-fraction field is determined by B; in the gas at $r = r_\ell$, $Z_\ell = B/(1 + B)$, while $Z = 1$ within the liquid, so that there is a discontinuity of Z at the interface. There is also a discontinuity of the fuel mass fraction since (3.18) gives $Y_{F\ell} = (B - C)/(1 + B) < 1$. Since $X_{F\ell} = Y_{F\ell} \bar{W}_\ell / W_F$, this last result determines the fuel mole fraction in the gas at the liquid surface and therefore the corresponding partial pressure, $pX_{F\ell}$. For evaporative equilibrium this determines the T_ℓ appearing in (3.15) for B as the boiling point at the subatmospheric pressure $pX_{F\ell}$. Iteration may thus be needed to evaluate B.

When the transfer number is known, the mass transfer rate is given by (3.16). Since the mass loss causes the droplet radius to decrease, if the liquid density ρ_ℓ remains constant it is found that

$$4\pi r_\ell^2 \rho_\ell \, dr_\ell / dt = -\dot{m}. \qquad (3.21)$$

If ρD_{12} and B remain constant during burning, then (3.16) enables (3.21) to be integrated to give

$$r_\ell^2 = r_{\ell_o}^2 - t\left[2\left(\rho D_{12}\right)/\rho_\ell\right] \ln\left(1 + B\right), \qquad (3.22)$$

where r_{ℓ_o} is the initial radius. The functional form of (3.22) is evident in advance from dimensional analysis, since the process is diffusion-controlled, and r_ℓ^2/t has the dimensions of a diffusion coefficient. This result is conventionally expressed as the "d-square law" of droplet burning; the square of the droplet diameter decreases linearly with time with a constant of proportionality given by the evaporation constant

$$K \equiv \left[8\left(\rho D_{12}\right)/\rho_\ell\right] \ln\left(1 + B\right). \qquad (3.23)$$

Since Lewis numbers are unity in the approximations that lead to (3.23), ρD_{12} can equally well be replaced by λ/c_p. When Lewis numbers are not unity it is better to use the latter to calculate K for droplet burning because the heat transfer is the dominant evaporation driver. In this respect, droplet burning differs from pure evaporation in a cool atmosphere, for example. Droplet-combustion investigations have found that λ/c_p between the fuel and the flame is most important to K, ρD_{12} for oxidizer-product diffusion is second most important, and ρD_{12} for fuel-product diffusion is least important (since changing it changes r_c/r_ℓ to keep K nearly fixed). In calculating λ/c_p, λ may be taken as an average for the fuel-product mixture at the average of the surface and flame temperatures, but c_p should be that for the fuel alone at this temperature because general theory shows that in a transfer process a species must be transported for its heat capacity to affect the transfer rate (Law and Williams, 1972).

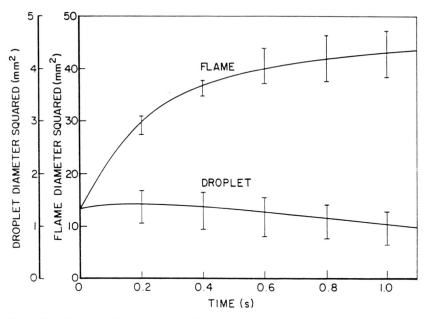

Fig. 3.10. Squares of droplet and flame diameters as functions of time for a representative test of decane in air.

There have been many experimental tests of the predictions of (3.20) and (3.22). An example for a decane droplet burning in room-temperature air at atmospheric pressure is shown in Fig. 3.10. These experiments were performed in a free-fall apparatus to circumvent buoyancy. After a heatup period it is seen that the d-square law is observed. The results are consistent with (3.22) and with the theoretical value $K = 0.6$ mm^2/s for these conditions. On the other hand, the flame diameter continually increases with time over the observation period, and (3.20) does not apply, for reasons indicated previously. In general, (3.20) usually is inaccurate even when the flame is in the quasisteady zone because of needed corrections for differing and variable transport coefficients.

Flame Spread

Mass-transfer theory is also useful for describing burning rates (linear regression rates normal to the burning surface) of fuels in fires (Williams, 1982). Also of interest in fires is the rate of flame spread through unignited fuel. The spread rate differs from the burning rate and under appropriate conditions (such as "crown fires" in the tops of trees in a dry forest) can be very rapid. A general approach to the description of flame-spread rates in complex configurations is first to identify a surface of fuel involvement that separates the burning fuel from that which is not burning (Williams, 1977). This surface, which flames approach at some but not all points, is the boundary of a control volume within which the virgin fuel is contained. An ignition stimulus, generally in the form of energy, must be transferred across the surface of fuel involvement to cause flame spread. An equation for the spread velocity V can be written as

$$\rho_F V \int\limits_{T_o}^{T_i} c_F dT = q,\qquad(3.24)$$

where ρ_F is the fuel density, c_F is its heat capacity, T_o its initial temperature and T_i its "ignition" temperature at which it becomes involved in flame, and q is the rate of energy transfer per unit area from the flames across the surface of fuel involvement. Equation (3.24) enables different physical processes of flame spread to be investigated in an approximate way, without attempting to solve differential conservation equations. This is especially helpful for complex processes in complicated configurations. Integral conservation principles are more useful than the differential equations for complicated problems. In a sense (perhaps paradoxically), the simpler a process is the more reasonable it is to attempt to obtain very accurate descriptions of it by solving complicated differential equations; for complicated processes accurate descriptions are unachievable, and it is more prudent to seek understanding through greatly simplified equations (Williams, 1977).

Deflagration propagation can be viewed as a flame-spread process. In this application (3.24) is a version of (2.3) and (2.4), but its use is unnecessary because much more accurate descriptions are available. A problem for which (3.24) is more appropriate is the spread of flames through a porous fuel bed such as a pile of sticks. In this application, the bed contains pockets of air, the heating of which does not contribute a resistance to flame spread. In (3.24) ρ_F therefore is $f\rho_s$, where ρ_s is the density of the solid fuel and f is the packing fraction, the ratio of the volume of the bed occupied by fuel to the total volume of the bed. For wood T_i will be its gasification temperature, about 650 K, the temperature at which rapid pyrolysis begins to occur to liberate combustible gases. If the fuel elements are thick, they are not heated completely before flame arrival, and ρ_F should be $fg\rho_s$, where g is the volume fraction of fuel heated. If the heating extends a distance h within the bed, then the heating time of an element is h/V, and the depth heated will be about $\sqrt{D_F h/V}$, where D_F is the thermal diffusivity of the fuel $\lambda_F/(\rho_s c_F)$, so that g will be $s\sqrt{D_F h/V}$, where s is the surface-to-volume ratio of the fuel elements, approximately the reciprocal of the diameter for long sticks. For radiant energy penetration into a porous bed, h is roughly $(fs)^{-1}$, so that (3.24) gives

$$V = q^2 \left[fs D_F \rho_s^2 \left(\int_{T_o}^{T_i} c_F dT \right)^2 \right]^{-1} , \qquad (3.25)$$

in which q is $\sigma\epsilon_f T_f^4$, where T_f is a flame temperature and ϵ_f a flame emissivity. Equation (3.25) shows how the spread velocity through the bed depends on the bed properties for "thermally thick" fuels. For "thermally thin" fuels, $g = 1$, and the alternative formula

$$V = q \left[f\rho_s \int_{T_o}^{T_i} c_F dT \right]^{-1} \qquad (3.26)$$

is obtained, which should be used instead of (3.25) if $s\sqrt{D_F h/V} > 1$.

For downward spread of a diffusion flame along the vertical surface of a thermally thick fuel sheet, q may occur mainly by heat conduction through the gas and geometrically is then approximately $(\ell/w)\lambda_G (T_f - T_i)/d$, where ℓ is the vertical distance along the virgin fuel surface over which the heat input occurs, w is the fuel thickness, and d is the average distance in the gas from the flame to the fuel surface, the subscript G identifying gas properties. Here g is about $\sqrt{D_F \ell/V}/w$, and with c_F taken constant and ℓ estimated as D_G/V_G, a heat-conduction length for downward penetration into a gas flowing upward at velocity V_G, (3.24) gives

$$V = V_G (\Gamma_G/\Gamma_F)^2 (T_f - T_i)^2 / (T_i - T_o)^2 , \qquad (3.27)$$

where the thermal responsivity (or thermal inertia) of a medium is $\Gamma = \sqrt{\rho c\lambda}$, and d has been put equal to ℓ. In (3.27) V_G may be estimated as a buoyant velocity constructed from the thermal diffusivity of the gas and the acceleration of gravity; in forced flow it would be the forced opposing gas velocity. For thermally thin fuels, again $g = 1$, and the simpler formula

$$V = [\lambda_G/(\rho_F c_F w)] (T_f - T_i)/(T_i - T_o) \qquad (3.28)$$

is obtained.

Ideas of this kind can be applied to many other configurations of flame spread. Application has been made to upward flame spread along vertical fuel surfaces, which has been reasoned to be accelerating under many conditions, to wind-aided spread over horizontal solid fuel surfaces, to buoyancy-driven spread along horizontal surfaces of liquid fuel pools, and to spread over liquid pools driven by surface-tension gradients, for example. Flame-spread processes often exhibit intricacies that make them one of the most interesting aspects of combustion. A difficult problem, on which progress has been made recently, is the spread of the edge of a diffusion flame into a mixing layer between gaseous fuel and oxidizer (Dold, 1988; Liñán, 1991).

Current Problems in Diffusion Flames

Table 3.3 is a list of some current combustion problems in the area of diffusion flames that require further research, most of the items in this list have been mentioned earlier in this chapter. The bibliography offers some entries into recent literature.

Table 3.3. Current problems in diffusion flames that need further study

1. Structures with real kinetics (hydrocarbon–air, H_2–O_2, H_2–halogen)
2. Extinction with chemical suppressants
3. Structures and extinctions with sprays or powders
4. Nonplanar instability
5. Interaction perpendicular to a cold wall
6. Opposed-flow spread over solids with chemical kinetics controlling
7. Spread over liquids with buoyancy and surface tension

Bibliography

Burke, S.P. and Schumann, T.E.W. 1928. *Ind. Eng. Chem.* **20**, 998.
Chelliah, H.K. and Williams, F.A. 1990. *Combust. Flame* **80**, 17.
Cho, S.Y., Choi, M.Y., and Dryer, F.L. 1991. *Twenty-third symposium (international) on combustion*. The Combustion Institute, Pittsburgh, pp. 1611–1617.
Damköhler, G. 1936. *Zeit. Elektrochem.* **42**, 846.
Dold, J.W. 1988. *Prog. Astronaut. Aeronaut.* **113**, 240.
Drew, D.A. 1983. *A. Rev. Fluid Mech.* **15**, 261.
Faeth, G.M. 1983. *Prog. Energy Combust. Sci.* **9**, 1.
Glassman, I. 1987. *Combustion*, 2nd ed. Academic Press, New York.
Howard, J.B. 1991. *Twenty-third symposium (international) on combustion*. The Combustion Institute, Pittsburgh, pp. 1107–1127.
Kanury, A.M. 1975. *Introduction to combustion phenomena*. Gordon and Breach, New York.
Law, C.K. and Williams, F.A. 1972. *Combust. Flame* **19**, 393.

Liñán, A. 1974. *Acta Astronautica* **1**, 1007.

Liñán, A. 1991. *El papel de la mecánica de fluidos en los procesos de combustión.* Real Academia de Ciencias Exactas, Fisicas y Naturales, Madrid.

Rosner, D.E. 1986. *Transport processes in chemically reacting flow systems.* Butterworths, New York.

Saito, K., Williams, F.A., and Gordon, A.S. 1986. *J. Heat Transfer* **108**, 640.

Seshadri, K., Mauss, F., Peters, N., and Warnatz, J. 1991. *Twenty-third symposium (international) on combustion.* The Combustion Institute, Pittsburgh, pp. 559–566.

Seshadri, K. and Peters, N. 1988. *Combust. Flame* **73**, 23.

Seshadri, K. and Peters, N. 1990. *Combust. Flame* **81**, 96.

Wallis, G.B. 1969. *One-dimensional two-phase flow.* McGraw-Hill, New York.

Williams, F.A. 1977. *Sixteenth symposium (international) on combustion.* The Combustion Institute, Pittsburgh, pp. 1281–94.

Williams, F.A. 1981. *Fire Saf. J.* **3**, 163.

Williams, F.A. 1982. *Prog. Energy Combust. Sci.* **8**, 317.

Williams, F.A. 1985. *Combustion theory*, 2nd ed. Addison-Wesley Publishing Company, Menlo Park, California.

4

FLAMMABILITY, EXPLOSIONS, AND DETONATIONS

There is both fundamental and practical interest in knowing what constitutes a combustible mixture capable of experiencing deflagration or detonation and in determining what it takes to ignite such a mixture or to make it produce the strong pressure pulses normally associated with explosions. These problems in flammability, ignition, and explosions are addressed in the present chapter. They involve the self-acceleratory character of combustion and the definition of critical conditions needed for self-acceleratory behavior to occur. Some of the experimental observations, as well as theoretical approaches that describe them, will be reviewed.

In addition, attention will be devoted to the detonations defined in Chapter 1. Consideration will be given to detonation structure, detonation stability, and conditions for initiation and failure of detonations. Thus, while pressure has been nearly constant in the combustion processes discussed in previous chapters, large pressure changes occur in many of the phenomena to be considered here. This is especially significant for effects of combustion on surrounding materials. However, it will be seen in the following sections that pressure changes are relatively small and often unimportant in the initial stages of explosions.

The Self-Acceleratory Nature of Combustion

Two different kinds of phenomena can contribute to acceleration of combustion processes—chain branching and thermal runaway. Most explosion processes in combustion involve both of these. By "acceleration" here it is meant that the rate of the combustion process increases as it proceeds. Separate discussions of chain branching (which leads to branched-chain explosions) and of thermal runaway (which leads to thermal explosions) help to clarify the concepts. Semenov (1935) had much to do with laying the foundations of the subject, and Gray and Lee (1967) provide a review of much of the early work.

In Chapter 2 it was seen that the rates of hydrogen and hydrocarbon combustion processes increase as the concentrations of the active species, H, OH, and O, increase. The elementary step $H + O_2 \rightarrow OH + O$ was cited as an example of a chain-branching step that increases the total concentration of these species (the concentration of the "radical pool"). As a simple illustration of the consequences of branching, let c represent chain-carrier concentration, and assume that the carrier is produced in an initiation step at a constant rate k_I, may multiply in a propagation step with rate proportional to c (the proportionality factor for propagation being a constant reciprocal-time k_P) and having a constant average branching b (the number of carrier produced per carrier consumed), and is consumed in a termination step with a rate also proportional to c (the proportionality factor for termination being a constant reciprocal-time k_T). Then, in a spatially homogeneous, constant-volume system, (1.4) gives the reaction-accumulation balance,

$$dc/dt = k_I + (b - 1)k_P c - k_T c , \qquad (4.1)$$

the integral of which is

$$c = c_o e^{[(b-1)k_P - k_T]t} + k_I \left\{ e^{[(b-1)k_P - k_T]t} - 1 \right\} / [(b-1)k_P - k_T] , \qquad (4.2)$$

where c_o is the concentration at time zero.

If $k_T > (b-1)k_P$, then (4.2) shows that for large times c approaches the constant value $k_I / [k_T - (b-1)k_P]$, but if $k_T < (b-1)k_P$, c grows exponentially with time for large time, as $\{c_o + k_I / [(b-1)k_P - k_T]\}$ $e^{[(b-1)k_P - k_T]t}$. This exponential multiplication of chain-carrier concentration is representative of self-acceleratory behavior and characteristic of branched-chain explosions; with no carriers initially present, the initiation step would produce them linearly, and after they build up to a concentration of about $k_I / [(b-1)k_P - k_T]$, the branching step would become dominant and cause the exponential growth. The condition for occurrence of a branched-chain explosion, that is, the condition for existence of the exponential growth, is

$$(b - 1)k_P > k_T , \qquad (4.3)$$

which states that the branching rate associated with propagation exceeds the removal rate through termination. Although real chemistry involves more than one carrier and nonlinear differential equations for each, the general character remains the same. In practice, often k_T/k_P is small, so that relatively small average branching can give explosion.

Thermal runaway is a consequence of energy conservation and the strong temperature dependence of the overall rate of heat release. It can be discussed on the basis of a one-step, Arrhenius approximation for the chemistry with influences of changes in species concentrations neglected. If the rate of heat release divided by the heat capacity is Ae^{-E/R_oT}, where A and E are constants, then in a spatially homogeneous system, energy conservation (1.3) gives

$$dT/dt = Ae^{-E/R_oT} \tag{4.4}$$

for the rate of increase of temperature. For small temperature changes about initial temperature T_o, the expansion $T^{-1} = T_o^{-1}[1-(T-T_o)/T_o + \cdots]$ reduces (4.4) to

$$dT/dt = \left(Ae^{-E/R_oT_o}\right) e^{(E/R_oT_o^2)(T-T_o)}, \tag{4.5}$$

the solution to which is

$$T = T_o + \left(\frac{R_oT_o^2}{E}\right) \ln\left[\frac{1}{1-(AE/R_oT_o^2)e^{-E/R_oT_o}t}\right]. \tag{4.6}$$

According to (4.6), the temperature increases at first linearly with time but then accelerates, approaching infinity at the finite time

$$t_i = \left[R_oT_o^2/(AE)\right]e^{E/R_oT_o}. \tag{4.7}$$

The infinite temperature is a result of the approximation (4.5); according to (4.4), the increase would decelerate and become linear when T becomes of order E/R_o, but this temperature is so high that (4.4) is invalid, and reactant depletion causes the temperature to level off earlier. Nevertheless, the approach to a very large rate of temperature increase at $t = t_i$ shows that the thermal explosion is much more precipitous than the branched-chain explosion because of the nonlinearity in the rate expression of (4.4).

For the thermal explosion, t_i of (4.7) is an estimate of the explosion time, the time required for the explosion to occur. For the branched-chain explosion, a corresponding estimate would be $[(b-1)k_P - k_T]^{-1}$. There is some arbitrariness in these estimates because the explosion processes are evolving in time. A more thorough treatment of a particular example of a thermal explosion will clarify these ideas further and will better identify critical conditions for explosions to occur.

Initiation of Thermal Explosions

A representative laboratory experiment on explosion is illustrated in Fig. 4.1. An evacuated chamber of volume V and surface-area S is heated to a wall temperature T_w. At time zero, a valve is opened to admit a cool reactant gas mixture to the chamber through a heating element that rapidly raises the gas temperature to T_w. When the desired chamber pressure is reached, the valve is closed, and the history of the reaction is observed, for example by thermocouples, by pressure gauges, or by gas sampling, and optically, while the wall temperature is maintained at T_w.

This experiment affords test times of about one second or more. For shorter test times a rapid-compressor machine may be employed, in which the cold reactant mixture, contained within a chamber, is heated adiabatically by moving a piston in a single stroke. The use of shock tubes (which have been of great importance in studies of the chemical kinetics of combustion) affords even shorter times, in the millisecond range. In general, the laboratory test method is selected to be consistent with the time scales of the phenomenon investigated.

As an approximate description of the experiment in Fig. 4.1, the gas may be treated as being spatially homogeneous, with influences of heat transfer to the wall approximated through a heat-transfer coefficient. Use of (1.4) in (1.3) and integration of (1.3) over the volume of the vessel shows that the rate of increase of the thermal energy of the gas is

$$V \rho c_v dT/dt = V \rho c_v A e^{-E/R_o T} - S(\lambda/\ell)(T - T_w) , \qquad (4.8)$$

where A is the same as in (4.4) since one-step, Arrhenius chemistry is considered with reactant depletion neglected. In the last term of (4.8),

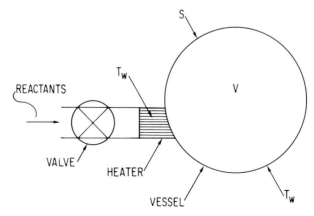

Fig. 4.1. Example of an experiment on homogeneous thermal explosions.

the ratio of the thermal conductivity λ to the characteristic length ℓ, times the temperature difference $(T - T_w)$, is an approximation to the rate of energy loss per unit wall area by heat conduction. Here (λ/ℓ) represents a heat-transfer coefficient, employed extensively in engineering heat-transfer analyses. If the gas is truly at rest, then ℓ is of the order of V/S but depends on the shape of the vessel. If the gas is in motion, the ℓ is a boundary-layer thickness, which is smaller. The initial condition for (4.8) is $T = T_w$ at $t = 0$.

Introduction of nondimensional variables aids in integration of (4.8), as well as in discussion of the results of numerical integrations. Studies have identified an efficient selection of variables. A characteristic heat-loss time (or cooling time) is

$$t_\ell = V \rho c_v / (S\lambda/\ell) , \qquad (4.9)$$

and appropriate nondimensional temperature and time are

$$\theta = (T - T_w)/(\epsilon T_w) , \tau = t/t_\ell , \qquad (4.10)$$

where a small parameter (typically of order 10^{-2}) is

$$\epsilon = RT_w/E . \qquad (4.11)$$

Substitution into (4.8) then gives

$$d\theta/d\tau = \delta e^{\theta/(1+\epsilon\theta)} - \theta , \qquad (4.12)$$

where δ, a parameter named after Frank-Kamenetskii because of his extensive research in the area (Frank-Kamenetskii, 1969), is

$$\delta = \left(\frac{E}{RT_w} \right) \left(\frac{At_\ell}{T_w} \right) e^{-E/R_o T_w} . \qquad (4.13)$$

In terms of the heat release Q per unit mass of the reactant mixture consumed, the nondimensional heat release $Q/(c_v T_w)$ occurs as a factor in (4.13) when a chemical time t_c for heat release at temperature T_w is introduced,

$$t_c = [Q/(c_v A)] e^{E/R_o T_w} . \qquad (4.14)$$

Aside from the factors $E/(R_o T_w)$ and $Q/(c_v T_w)$, it is seen that δ is the ratio (t_ℓ/t_c) of a transport heat-loss time to a chemical heat-release time. For small values of $\epsilon\theta$, the term $\epsilon\theta$ may be neglected in (4.12), giving a one-parameter problem, with the single parameter δ; the initial condition is $\theta = 0$ at $\tau = 0$.

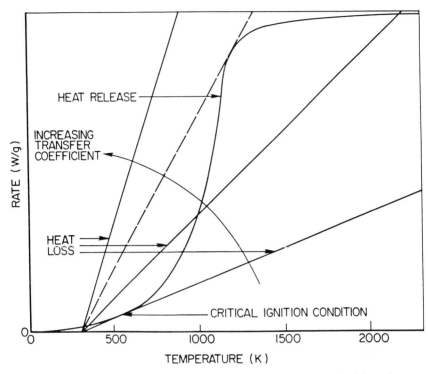

Fig. 4.2. Schematic dependences of rates of heat release and of heat loss on temperature, illustrating criticality.

The character of the solution can be seen by first investigating conditions for existence of a steady state, $d\theta/d\tau = 0$. At a steady state, the equation

$$\theta/\delta = e^{\theta/(1+\epsilon\theta)} \approx e^\theta \qquad (4.15)$$

must be satisfied. The existence of solutions depends on the ratio of the heat-loss rate to the heat-release rate, δ^{-1}. Straight lines representing the loss rate and a curve representing the generation rate are shown schematically in Fig. 4.2. As seen in Fig. 4.2, for small heat-transfer coefficients (small δ^{-1}), there is no solution except at unreasonably high temperatures ($\theta = \delta^{-1}e^{\epsilon^{-1}}$ according to the previous equation), and the heat-generation rate exceeds the loss rate for all reasonable θ. Under these conditions the loss can be neglected, and the problem (4.4) is recovered, with the approximate solution (4.6); a thermal explosion occurs. At larger values of δ^{-1}, Fig 4.2 shows two intersections defining steady states (three, if $\epsilon\theta$ is not neglected). The lower intersection is stable, as seen from the fact that if the temperature is increased, the loss rate exceeds the generation rate, so that the temperature tends to decrease back to the steady state. On the contrary, the second intersection is unstable and therefore not a realizable steady state. The uppermost intersection is stable but not meaningful unless reactant depletion is included; it would represent a rapidly burning condition, and increasing the loss rate until this intersection no longer exists (the dashed curve in Fig. 4.2) corresponds roughly to causing extinction, as discussed in the previous chapter for diffusion flames. The lower, stable intersection is that which is reached from the initial state $\theta = 0$ and represents continuing slow, nonexplosive combustion (an initial temperature above that of the second intersection would be needed for explosion to occur at this value of δ). The limit condition for nonexistence of the lower intersection, the value of δ^{-1} for the critical ignition condition shown in Fig. 4.2, defines the heat loss needed to just prevent the thermal explosion from occurring. With the approximation in (4.15), this is

$$\delta^{-1} \equiv \delta_c^{-1} = e \, , \tag{4.16}$$

which specifies the critical value of the Frank-Kamenetskii parameter.

The complete histories of solutions to (4.12) with $\epsilon \theta$ neglected are illustrated in Fig. 4.3. At large δ, thermal explosion occurs, with runaway at $\tau = \delta^{-1}$, which is a nondimensional statement of (4.7), applicable for highly supercritical conditions. At small, subcritical values of δ, the steady-state, slow reaction at $\theta = \theta_m(\delta)$ is approached as $\tau \to \infty$. For

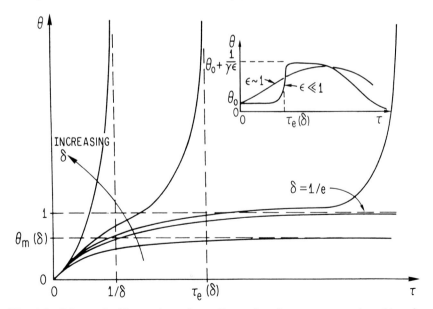

Fig. 4.3. Schematic illustration of nondimensional temperature-time histories for thermal explosions for the simplified Frank-Kamenetskii model (main graph) and for a more complete model that includes reactant depletion (inset).

slightly supercritical conditions, $0 < \delta - 1/e << 1$, θ remains near unity for a long time before the abrupt runaway occurs; this behavior has been described by asymptotic analysis that recognizes different stages (Kassoy and Liñán, 1978). The longer-time, supercritical history in the presence of reactant depletion is sketched in the inset in Fig.4.3, where the nondimensional initial temperature is θ_o and the nondimensional adiabatic flame temperature is $\theta_o + 1/(\gamma\epsilon)$; the abrupt runaway is seen there to occur only if $\epsilon << 1$, while if ϵ were of order unity, a gradual evolution of temperature would be observed, with no distinguished time of an explosion event.

Since the parameter δ takes on a particular value at criticality, studies have been pursued to find how this value depends on the shape of the vessel. Initial-boundary value problems for partial differential equations have been considered in which the differential form of the heat-conduction term is retained. Since gas motion usually is not included in these studies, they apply better to thermal explosion of solid reactants than to gaseous combustibles. Complete results are available for spherical, cylindrical, and planar (slab) geometries, but influences of departures from these simple symmetries continue to pose challenging questions. Values of δ_c for the slab, the cylinder, and the sphere are available, and they decrease in that order, as expected from the greater ease of heat loss in a larger number of dimensions. It is found again that the intersections at the higher temperatures are unstable, so that Fig. 4.2 provides a correct qualitative description. Interesting mathematical peculiarities arise in higher dimensions, such as a large number of steady states at higher temperatures, which however are not of physical significance (Williams, 1985).

Applications of thermal explosion theory arise not only for reactant gas mixtures in vessels but also for solid reactants. There are exothermic solids, such as NH_4NO_3, a component of fertilizer, that liberate heat by finite-rate chemistry. In view of (4.9), according to (4.13) δ increases as the size of the system increases. Therefore if exothermic solids are stored in piles that are too large, δ may exceed δ_c, and thermal explosion will occur. Fertilizer transported on barges has been observed to explode if packed in piles that are too large. Therefore there are safety rules in transport that specify maximum pile dimensions. "Spontaneous combustion" is another name for the process under discussion. For example, piles of oily rags can spontaneously burst into flame by a process like that described here. In this last example, the term "explosion" usually would not be employed because achievement of criticality results in a diffusion-flame form of combustion.

The term "explosion" is derived from damaging overpressures that may build up after criticality. Description of the pressure field necessitates inclusion of the full set of conservation equations, particularly mass and momentum conservation, (1.1) and (1.2), and allowance for spatial nonhomogeneity. Theoretical and experimental studies of these destructive stages of thermal explosions have been pursued. Depending on conditions, a variety of different phenomena may be encountered. For the experiment of Fig. 4.1, in a spherical chamber the hottest spot may be expected to occur in the center, and after runaway a moving ignition front leading to deflagration may develop there and propagate toward the wall. The size of the hot spot may increase or decrease with time during different stages of the process (Dold, 1989). Pressure waves are generated from the hot spot and the deflagration (Jackson, Kapila, and Stewart, 1989; Kassoy, Kapila, and Stewart, 1989) and may coalesce into shock waves, undergoing a transition to detonation in extreme cases. For very large chambers, there can be self-similar stages, in which the explosion does not develop into detonation but instead the deflagration drives a shock wave far ahead of it. Although much information is available, the problems in developing a full understanding of the process are so difficult that they will provide a source of research for chemical kineticists, gas dynamicists, and applied mathematicians for many more years. Two relevant recent reviews are those of Kassoy (1985) and of Choi and Majda (1989).

Ignition

In the processes considered above, the explosion developed spontaneously, without application of an ignition source. The later stages involving pressure waves can be similar to those described irrespective of the manner of ignition, but introduction of an ignition source can cause explosion in systems that would not experience it spontaneously. Ignition can lead to explosion or merely to deflagration without appreciable pressure pulses, depending on the situation. There are a wide variety of methods to achieve ignition. These include exposure of the combustible to a hot solid or gas, exposure to a flame (a pilot), application of radiant energy input, discharge of an electrical spark in the combustible, and initiation of an explosive charge that communicates with the combustible. Criteria for achieving ignition are expressible as spontaneous ignition temperatures for low rates of energy input and as minimum ignition energies for high rates of energy input. Values of these quantities for various fuels in air (Williams, 1987) are listed in

Table 4.1. Ignition and flammability properties of selected fuels in air

Fuel	Minimum ignition energy (mJ)	Spontaneous ignition temperature (K)	Lower flammability limit (% by volume of gaseous fuel in mixture)	Upper flammability limit (% by volume of gaseous fuel in mixture)	Minimum quenching distance (cm)	Maximum burning velocity (cm/sec)
H_2	0.02	850	4.0	75.0	0.06	300
CO	—	900	12.0	74.0	—	45
CH_4	0.29	810	5.0	15.0	0.21	44
C_2H_6	0.24	790	3.0	12.0	0.18	47
C_2H_4	0.09	760	2.7	36.0	0.12	78
C_2H_2	0.03	580	2.5	100.0	0.07	160
C_3H_8	0.24	730	2.1	9.5	0.18	45
C_7H_{16}	0.24	500	1.1	6.7	0.18	42
$C_{12}H_{26}$	—	480	0.6	—	0.18	40
C_6H_6	0.21	840	1.3	8.0	0.18	47
CH_3OH	0.14	660	6.7	36.0	0.15	54

Table 4.1 (which also gives maximum values of the laminar burning velocity, and other flammability properties, to be discussed later). The values vary strongly with the equivalence ratio, for example, as well as with experimental configurations; minimum values are listed in the table. Since ignition temperatures are achieved by slow energy input, an alternative long-time ignition criterion would be the minimum rate of energy addition (minimum ignition power) needed to maintain the thermal energy at a level corresponding to the ignition temperature. This alternative criterion, although of more direct operational utility, varies more strongly with the ignition stimulus and with the configuration and environment of the combustible.

Ignition processes involve both space and time dependences, as well as chemical heat release, and therefore necessitate the use of nonlinear partial differential equations for describing them. Typically they exhibit distinct stages in the history of the process and within each stage, different zones maintaining balances of different effects. As an example, the ignition of an exothermic solid by a constant, externally applied, radiant energy flux may be considered (Liñán and Williams, 1971).

If the semi-infinite solid is assumed to occupy the region $x > 0$ and the constant energy flux q is applied at the surface $x = 0$, starting at time $t = 0$, then the problem in its simplest form can be written as

$$\rho c \partial T / \partial t - \lambda \partial^2 T / \partial x^2 = \rho c A e^{-E/R_o T} \qquad (4.17)$$

for $0 < x < \infty$, $0 < t < \infty$, with $T = T_o$ (the initial temperature) at $t = 0$ for all $x \geq 0$ and with

$$\lambda \partial T / \partial x = -q \qquad (4.18)$$

at $x = 0$ for all $t > 0$. Here c is the constant heat capacity of the solid, ρ its constant density, λ its constant thermal conductivity, and A the constant prefactor of (4.4), in the one-step, Arrhenius approximation with reactant depletion neglected. The presence of the heat flux shortens the time of thermal runaway significantly below that given by (4.7) because it raises the surface temperature above T_o. This problem has been solved by both numerical integration and asymptotic expansion in the small parameter $R T_c^2 / (E T_o \sqrt{\pi \tau_c})$, where the nondimensional time τ is $q^2 t / (\rho \lambda c T_o^2)$, its critical value τ_c is related to T_c by

$$T_c = T_o (1 + 2\sqrt{\tau_c / \pi}) , \qquad (4.19)$$

and the first approximation for the critical temperature T_c is

$$T_c = \frac{E}{R_o} \left[\ln \left(\frac{\lambda \rho c T_o A}{q^2} \right) \right]^{-1} . \qquad (4.20)$$

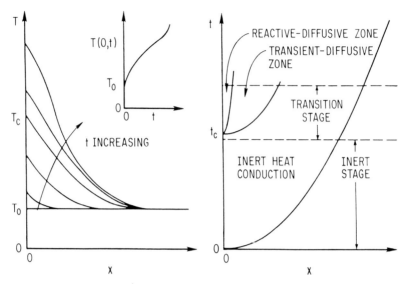

Fig. 4.4. Illustration of the solution to the problem of ignition by a constant heat flux.

Both methods show ignition to occur at a critical time t_c given by

$$t_c = \left(\frac{\pi\lambda\rho c T_o^2}{4q^2}\right)\left\{\left[\ln\left(\frac{\lambda\rho c T_c A(\pi\tau_c)^{1/4}}{0.65q^2\sqrt{E/R_oT_o}}\right)\right]^{-1}\left(\frac{E}{R_oT_o}\right) - 1\right\}^2,$$

(4.21)

in which (4.19) and the definition of τ are to be employed inside the logarithm in an iterative calculation for obtaining the most accurate results. The history of the process is illustrated in Fig. 4.4.

It is found that there is an initial stage of inert heat conduction, in which a heated layer develops at the surface and spreads into the interior at a rate illustrated in the space-time plot at the right of Fig. 4.4. The reaction term is negligible during this stage. Near time t_c, this stage ends, and a stage of transition to ignition begins. In this transition stage, the reaction term is important at the heated surface. There is a reactive-diffusive zone adjacent to the surface, where the transient (accumulation) term is negligible and a transient-diffusive zone next to this, where the reaction term is negligible. Both of these zones are thin compared with the total thickness of the heated layer at time t_c. Here t_c is an example of an ignition delay time—the time from application of an ignition stimulus until ignition occurs. Many ignition processes of related types have been analyzed and now are understood reasonably well (Williams, 1985).

Flammability and Explosion Limits

In the experiment of Fig. 4.1, the critical temperature needed for explosion to occur can be measured as a function of the pressure of the combustible gas mixture to give an explosion-limit curve. From (4.9) and (4.13) it would be anticipated that this temperature should increase as the pressure decreases, to roughly correspond to a constant rate of heat release. Experimental results usually exhibit this general trend but often also show marked deviations from it. Measured explosion-limit curves for a particular hydrogen-oxygen mixture (dashed curve) and for a particular propane-oxygen mixture (solid curves) are shown in Fig. 4.5. At high pressures the hydrogen-oxygen curve has the expected trend for thermal explosion, but below about 0.2 atm it shows an "explosion peninsula" within which explosion occurs at a much lower temperature than would be anticipated from the high-pressure part of the curve. There is an appreciable range of temperature over which no explosion occurs if the pressure is low enough or high enough, but in an

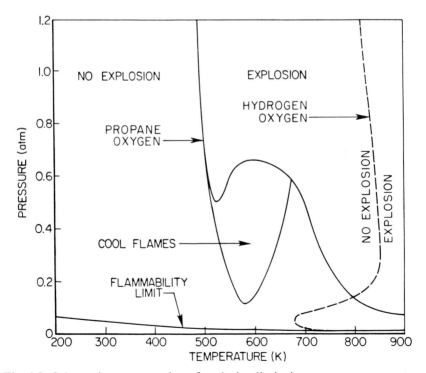

Fig. 4.5. Schematic representation of explosion limits in a pressure-temperature diagram for C_3H_8-O_2 mixtures having 50% fuel (solid curves) and for H_2-O_2 mixtures having 67% fuel (dashed curve).

intermediate pressure range explosion is observed. The chemical kinetics of branched-chain explosions produce this distinctive behavior.

In (4.3), which defines conditions for branched-chain explosions, the ratio k_T/k_P can depend nonmonotonically on pressure. Propagation steps are bimolecular, but there are two kinds of termination steps—three-body recombinations in the gas, and wall recombinations resulting from diffusion of radicals to the walls. Thus k_T/k_P has two terms, one proportional to p and the other proportional to p^{-2} (the additional power coming from the inverse proportionality of the diffusion coefficient to pressure). The three-body recombination makes (4.3) invalid at high pressures, and the wall recombination makes (4.3) invalid at low pressures, so that the explosion occurs in the intermediate pressure range. The dominant feature of the hydrogen-oxygen curve of Fig. 4.5 therefore is explained through ideas of branched-chain explosions. The high-pressure portion of the curve (called the "third explosion limit") may be influenced by the relatively stable HO_2 molecule's beginning to participate as a chain carrier, as well as by thermal-explosion phenomena. From these explanations it is expected that the explosion-limit curves will vary with vessel size, wall material, geometrical configuration, and so on. Experimentally, in fact, they do; differences of more than 100 K in the limit temperature at a given pressure are not uncommon.

The upper solid curve in Fig. 4.5 is the explosion-limit curve for the propane-oxygen mixture. Compared to the hydrogen-oxygen curve, it is somewhat closer to the thermal-explosion prediction but also exhibits notable nonmonotonicity, now of a different kind. This, too, is chemical-kinetic in origin. There is a region, identified in the figure as "cool flames," in which a few blue flames are observed to propagate sequentially through the mixture, without causing ignition or explosion. These cool-flame phenomena stem from particular chemical-kinetic effects that enable increases in intermediate concentrations to turn off the reaction after it has started, before appreciable depletion of the fuel has occurred. Although there has been much study of these phenomena, many aspects are poorly understood; research continues, and there is much more to be done.

The lowest solid curve in Fig. 4.5 is the flammability limit curve for the propane-oxygen mixture. In the region between this curve and the explosion-limit curve, although the mixture in the vessel does not explode of its own accord, it nevertheless is flammable in that, if an ignition stimulus such as a strong enough spark is applied, a deflagration can be initiated in the mixture. In the region below the flammability-limit curve, ignition cannot be achieved. The flammability-limit curve varies

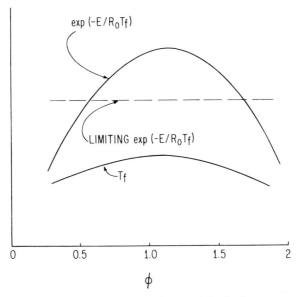

Fig. 4.6. Illustration of the dependence of the adiabatic flame temperature and of the Arrhenius factor on the equivalence ratio, exhibiting extinction condition for defining flammability limits.

strongly with the mixture ratio. At any given pressure and temperature, mixtures that are too lean or too rich in fuel are not flammable. These two limits are called the lower and upper flammability limits, respectively, and are listed in Table 4.1 for various fuels in room-temperature air at normal atmospheric pressure. As the pressure is decreased, the limits usually narrow and meet each other at a critical pressure, the minimum pressure limit of flammability.

Flammability limits (like explosion limits) depend on the experimental configuration. The values in Table 4.1 are obtained mainly from experiments in which a closed vertical tube 100 cm long and 5 cm in diameter is filled with a quiescent gas mixture that is subjected to a strong spark at the bottom; the mixture is judged flammable if a flame propagates all the way to the top. The flammable mixture-fraction range usually is wider for upward propagation than for downward propagation and wider in closed tubes than in open tubes, and it tends to increase as the size of the system increases. However, the 5-cm tube usually is large enough that the variation of the limits with size produced by a further increase in diameter are small. Theories for flammability limits consider deflagration propagation with heat loss and identify a critical ratio of the rate of heat generation to the rate of heat loss for extinction (that is, for the propagation limit to occur), as has been discussed in

Chapter 2. Since a roughly fixed rate of heat loss may be expected in a given configuration, the strong variation of the overall heat-generation rate with the equivalence ratio can be responsible for the lower and upper limits, as illustrated in Fig. 4.6. The steepness of this generation curve in the wings can be responsible for the weak dependences of the limits on configuration. It is equally possible that overall rate parameters, such as E, change abruptly at particular equivalence ratios for chemical-kinetic reasons, so that the kinetics exert a much stronger influence on the limits than the loss. Unsorting the relative importance of chemical-kinetic mechanisms and heat-loss rates on flammability limits remains an active and challenging area of research.

Entries in the penultimate column in Table 4.1, the minimum quenching distance, refer to conditions necessary to extinguish deflagrations. A propagating deflagration that encounters too narrow a restriction cannot penetrate to the other side; this fact underlies the design of flame arrestors in which fine grids are placed in gas lines to prevent flames from propagating through them. The quenching distance is the distance between a pair of parallel plates, dividing conditions for which a flame can propagate through a combustible mixture between them from conditions for which it cannot. The value is roughly representative of hole sizes (quenching diameters) needed to prevent flame penetration. Since the quenching is through heat loss, the value varies strongly with the mixture ratio, for the reason illustrated in Fig. 4.6. The minimum value is listed in Table 4.1, generally corresponding to near-stoichiometric mixtures. There are correlations between quenching distances and minimum ignition energies (Lewis and von Elbe, 1987); to ignite a combustible, an amount of energy must be deposited locally that is roughly sufficient to raise the temperature to the adiabatic flame temperature in a volume with a cross-sectional area of the order of the square of the quenching distance and with a thickness of the order of the laminar-flame thickness. Quenching distances typically are 20 to 60 times the laminar flame thickness because the strong temperature dependence of the heat-release rate enables a loss rate of about 10% of the release rate to extinguish the flame. While qualitative understandings of these flammability phenomena are available, there is much room for improvement in our detailed knowledge of the processes.

Detonation Structure

Since the most damaging form of explosion is a detonation, it is important to know the structures of detonations and the conditions for their occurrence. As a starting point for studying these questions, it is necessary to understand the structures of steady, planar detonations.

Detonations were defined in Chapter 1, where the jump conditions across them were derived; planar detonation structure differs greatly from deflagration structure, as illustrated in Fig. 4.7. While the steepest changes usually occur toward the back of a deflagration, they occur at the front of a detonation. Heat conduction and diffusion, which are important processes in deflagrations, are negligible in detonations because the gas velocities are so high. The heating of the gas is produced by a strong shock wave at the leading edge of a detonation, not by heat conduction. Across the shock, the pressure and temperature increase, the velocity decreases, and the chemical composition remains unchanged, at least in gaseous detonations, since there are too few collisions for chemical changes to occur; the shock is much thinner than illustrated in Fig. 4.7, where its scale has been expanded to exhibit changes more clearly. The chemistry generally occurs in a longer zone downstream from the shock, in a convective-reactive balance. Typically the kinetics are such that these changes occur more slowly at first (in an "induction zone"), then more rapidly, as illustrated. In the subsonic flow behind the shock, the heat release associated with the chemistry causes the pressure

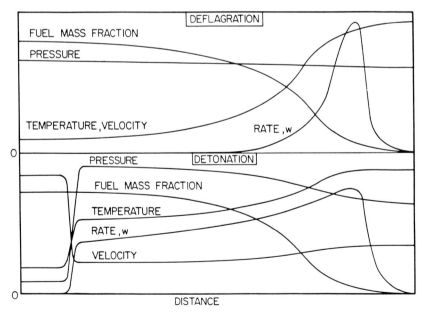

Fig. 4.7. Schematic illustration of representative structures of planar deflagrations and detonations.

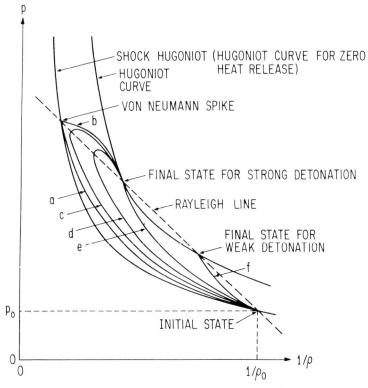

Fig. 4.8. Schematic illustration of detonation structure in a pressure-volume plane.

to decrease and the velocity and temperature to increase, as shown in Fig. 4.7.

The trajectory of the detonation structure can be followed on a pressure-volume diagram like that of Fig. 1.4. Such a diagram, drawn for gaseous detonations and containing only the detonation branch, is shown in Fig. 4.8. Two Hugoniot curves are shown, one for chemically frozen reactants and the other for equilibrium products. The structure that has just been described is curve *a* followed by curve *b*. Curve *a* is the shock structure in the frozen gas; it differs from both the Hugoniot and the Rayleigh line, being described, in fact, not by continuum equations but instead by kinetic theory, since these strong shocks are only a few molecular mean-free-paths in thickness (Vincenti and Kruger, 1965). The heat-release process, along curve *b*, follows the Rayleigh line to the extent that molecular transport is negligible (a good approximation); the departure from the Rayleigh line is exaggerated in Fig. 4.8 for illustrative purposes. It is seen that where curve *a* meets curve *b* is the highest pressure, the "von Neumann spike," which is roughly twice the final pressure, which itself can typically be 20 times the initial pressure. The structure (a, b) is called the "ZND structure" because it was found independently by Zel'dovich (1940), von Neumann (1942), and Döring (1943). Early experiments on detonations were performed by Berthelot and Vieille (1881) and by Mallard and le Chatelier (1881), and the jump conditions across them were derived by Chapman (1899) and Jouguet (1905). Reviews on detonation structure and dynamics are given by Zel'dovich and Kompaneets (1960), by Lee (1984), and by Strehlow (1984).

The structure (a, b) in Fig. 4.8 ends on the upper part of the equilibrium Hugoniot and therefore corresponds to a strong detonation. Since departures from the Rayleigh line through molecular transport can be shown to be in the direction of increased pressure, there is no way for curve b to reach the weak detonation branch without going past the equilibrium curve and ending in an endothermic process. Therefore only strong and Chapman-Jouguet detonations can possess the ZND structure. Real detonations do not quite reach the von Neumann spike, but they come much closer to it than curve c of Fig. 4.8. Curves d and e are hypothetical strong-detonation structures that do not approach the spike; chemical reaction rates are too slow for these structures to be seen in practice. Curve f is the only way to reach the weak-detonation end state. Weak detonations do not occur in combustion but have been found in certain flows, such as "condensation shocks" in wind tunnels, where water condenses in supersonic flow. The flow remains supersonic everywhere through these weak detonations.

In the strong and Chapman-Jouguet detonations, the flow behind the strong leading shock is subsonic. The Mach number increases as the heat is released, but does not reach unity for the strong detonation. For the Chapman-Jouguet wave, the velocity increases to sonic at the end of the detonation. It is mainly for this reason that detonations propagating steadily in tubes usually are Chapman-Jouguet waves; these detonations are followed by an unsteady expansion wave because of the dynamics of the flow in the tube, and this expansion can reach the strong detonation and interact with it to weaken it, but because of the sonic velocity, the expansion cannot reach the Chapman-Jouguet detonation to communicate with it. Strong detonations are observed in shock tubes with long driver sections that prevent the expansion wave from reaching the detonation during the test time and in standing detonations that can be produced in supersonic nozzle flow of combustible gases by suitably adjusting exit shock patterns.

Since transport properties are unimportant behind the leading shock of a detonation, accurate calculations of steady, planar, gaseous detonation structures with detailed chemical kinetics can be made relatively easily. Simplified approximations to these structures are helpful for studying the stability of the waves, for example. A useful approximation that needs further development is one in which there is an induction length ℓ_i behind the shock, over which no heat release occurs, followed by a short region of nearly instantaneous heat release. The distance ℓ_i is the induction time t_i times the velocity of the gas at the von Neumann state in a frame of reference fixed with respect to the detonation. For strong shock waves, in terms of the ratio γ of specific heats, this velocity is $v_o(\gamma - 1)/(\gamma + 1)$, where the detonation propagation velocity is approximately

$$v_o = \sqrt{2(\gamma^2 - 1)Q} \tag{4.22}$$

for Chapman-Jouguet waves, in which Q is the heat released per unit mass of the mixture. In terms of the final temperature T_∞ and molecular weight W_∞, the v_o of (4.22) can be shown to be approximately

$$v_o = \sqrt{2\gamma(\gamma + 1)T_\infty R_o/W_\infty} \tag{4.23}$$

and is found to lie between about 10^3 and 4×10^3 m/s; for stoichiometric H_2-O_2 mixtures it is 2830 m/s, for example, and for stoichiometric CH_4-O_2 mixtures, about 2500 m/s. With these results, the formula

$$\ell_i = t_i v_o(\gamma - 1)/(\gamma + 1) \tag{4.24}$$

may be employed, with t_i evaluated at the pressure and temperature of the von Neumann state. In these same approximations, the pressure and temperature needed for evaluating t_i are

$$p_i = p_o M_o^2(2\gamma)/(\gamma + 1) \tag{4.25}$$

and

$$T_i = T_\infty 2(\gamma - 1)/\gamma \tag{4.26}$$

where M_o is the propagation Mach number, typically between 5 and 10. The Mach number at the von Neumann state is about $\sqrt{(\gamma - 1)/(2\gamma)}$. Values of ℓ_i from (4.24) vary strongly with the equivalence ratio and decrease with increasing pressure; for gas mixtures initially at room temperature and atmospheric pressure, thicknesses in the range of 10^{-3} to 1 cm are representative.

Detonation Stability

For most real gaseous detonations, the ZND structure is made unstable by a mechanism that can be understood on the basis of the preceding simplified model. Exceptions are highly overdriven strong detonations with small heat release, which do not exhibit instability. The instability involves an acoustic-convective interaction, driven by the heat release at the end of the induction zone.

Suppose that a perturbation briefly introduces an increase in the propagation velocity v_o. Associated with this is an increase in the velocity v_i at the von Neumann state, and through the shock strengthening, a corresponding increase in T_i and p_i. According to (4.24), the associated fractional increase in ℓ_i is

$$\frac{\delta \ell_i}{\ell_i} = \frac{\delta v_i}{v_i} + \frac{\delta t_i}{t_i} = \frac{\delta v_o}{v_o} - \frac{E}{R_o T_i} \frac{\delta T_i}{T_i} - (n-1) \frac{\delta p_i}{p_i}$$

$$= \frac{\delta v_o}{v_o} \left[1 - 2\frac{E}{R_o T_i} - 2(n-1) \right] , \qquad (4.27)$$

where t_i has been assumed to respond to changes in T_i and p_i in accordance with a one-step, Arrhenius approximation with an overall reaction order (pressure exponent) n. The last equality in (4.27) is obtained by relating δp_i and δT_i to δv_o through (4.25) and a corresponding expression for T_i. When this perturbation, introduced at the shock, reaches the heat-release zone by convection, it causes a heat-release perturbation. An increase in the rate of heat release that causes ℓ_i to decrease will, in effect, generate a pressure pulse that can propagate upstream acoustically, further increasing v_o, and thus providing an instability mechanism. The condition for existence of this acoustic-convective instability is

$$\delta \ell_i / \delta v_o < 0 , \qquad (4.28)$$

which from (4.27) is seen to be

$$E/(R_o T_i) + n - 1 > 1/2 . \qquad (4.29)$$

The condition (4.29) is satisfied for most overall approximations to the real chemistry. In addition, stability calculations with detailed chemistry have exhibited a similar pulsating instability.

Since the instability involves downstream convection followed by upstream acoustic-wave propagation, its period is readily estimated as

$$t = \frac{\ell_i}{v_i} + \frac{\ell_i}{a_i - v_i} = \frac{t_i}{1 - M_i} \approx 10\frac{\ell_i}{v_o} , \qquad (4.30)$$

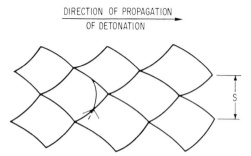

Fig. 4.9. Sketch of a regular pattern of smoked-foil inscriptions on the wall of a detonation tube with a rectangular cross section.

where a_i is the sound speed at the von Neumann state and M_i the Mach number there. Periods on this order have been observed for certain pulsating ("galloping") detonations. However, nonplanar stability analyses show further that this planar instability usually does not represent the most unstable mode. Instabilities involving transversely moving acoustic waves are stronger. These transverse waves, propagating at a preferred angle, give rise to the cellular detonation structure illustrated in Fig. 4.9. In this structure there are reactive triple-shock interactions, with the highest pressures and temperatures occurring where the three shocks meet. The trajectories of these triple points have been seen by high-speed spark-schlieren photography but can be observed much more easily by coating the walls of the tube with soot. After the detonation has propagated through the tube, a pattern of marks is left in the soot, where the triple points have eroded it away. The transverse cell spacing s, indicated in Fig. 4.9, can be measured from these patterns and is observed to be roughly 10 to 30 times ℓ_i. Although there have been some successful calculations of s through numerical integrations, the problem of predicting transverse spacings is difficult. Although there are a variety of important new ideas on this subject (stemming from bifurcation theory and other backgrounds), there is much more research to be done. Strehlow (1984) and Choi and Majda (1989) give relevant reviews.

The cellular structure certainly is a complicating aspect of detonation theory. Moreover, it is an aspect that was not widely recognized until the 1950s. Examples had been known at least since the 1920s of spinning detonations in which gaseous mixtures near the limit of combustion in tubes of circular cross section were observed to propagate in a helical path around the wall of the tube. At that time it was not realized that these are one-cell versions of cellular detonations. A systematic program of increasing the pressure or enriching the mixture to move away from the limit was not undertaken, even though the smoked-wall technique was available for observing two-cell, three-cell, etc., detonations. Instead, it was assumed that the cells would disappear when the mixture became detonable enough. The discovery therefore awaited modern fast-response (μ sec) instrumentation with high spatial resolution, which first identified a "turbulent" structure of healthy detonations in tubes, that only later was recognized to be the highly ordered cellular structure. A systematic scientific study could have revealed the phenomenon thirty years earlier.

The axial rate of progress of a spinning detonation down a tube is less than the propagation velocity of a planar Chapman-Jouguet wave. A fairly successful acoustic-interaction theory was developed to predict this difference (see, for example, Strehlow, 1984). In a healthy mixture with many cells, some parts travel faster than the planar ZND wave and others slower, so that the overall detonation velocity is essentially the Chapman-Jouguet velocity. Thus, by averaging over the cells, the planar theory can still be used to calculate the overall behavior, but not the detailed structure. As failure is approached, the single-cell behavior becomes more prominent, and transverse waves need to be considered. Whether this is also true for detonation of solids remains an open question. Investigation of whether cellular structure occurs for solid explosives is much more difficult because the sizes are much smaller and the pressures much higher.

Detonation Development and Failure

The transition from deflagration to detonation is a complicated process. The two different waves correspond to quite different combustion regimes in every respect; even the chemical kinetics differ because active intermediaries cannot diffuse into the fresh combustible to initiate the chemistry in the detonation. A dissociative initiation, such as $CH_4 + M \rightarrow CH_3 + H + M$, must be of importance in detonation. For a gaseous combustible in a tube, a flame can experience a transition to detonation only through an unsteady or nonplanar process, since Fig. 1.4 shows that there are no steady intermediate waves. Typically, expansion of the gases in the tube accelerates the flame and causes it to become turbulent, as well as emitting pressure waves that propagate in the direction of the deflagration. The pressure waves compress the fresh mixture and raise its temperature. The temperature increase can be great enough at positions of shock convergence to generate local explosions that develop into new propagating deflagrations moving outward from points often located at walls. From these ignition sites, deflagrations can travel in both directions along the tube, increasing the overall rate of

pressure-wave generation. Eventually, pressures in excess of final detonation pressures develop, and a detonation then propagates through the
rest of the combustible in the tube.

This sequence is favored by confinement because the walls reflect
the pressure waves back into the combustible. Openings allow relief
by transmitting pressure waves out of the system. In open combustible
clouds, transition to detonation is rare; although it can occur for sufficiently reactive combustibles, the time to transition is much longer.
Since partial confinement aids detonation development, studies of influences of partial confinement are important from the viewpoint of
safety.

In addition to developing from deflagrations, detonations can be
initiated directly by rapidly depositing a sufficient amount of energy in
the combustible, for example from a sufficiently large explosive charge.
Direct initiation is employed for liquid and solid explosives; sensitiv-
ity ratings of explosives are provided by tests in which the explosives
are subjected to successively stronger shock waves in standard configurations. In principle, direct initiation is easier to analyze than transition, although there still is an initial period of adjustment during which
tricky chemical-kinetic and fluid-dynamic details can affect whether the
detonation occurs. An approximate criterion for achieving direct initiation by instantaneous energy deposition is that enough energy must
be provided to maintain a shock as strong as the leading shock of the
Chapman-Jouguet detonation for a time as long as the chemical reaction
time of the steady, planar detonation.

Transmission of a detonation from a tube to an open cloud of the
same combustible is another process that has received recent attention
(Lee, 1984). It has been found that transmission occurs if the tube diameter exceeds $13s$, about 200 ℓ_i, where s and ℓ_i are the cell spacing
and induction length defined previously. Reasons for this have been
proposed but need further elucidation. The problems are in gas dynamics of reactive systems with negligible molecular transport but strong
shocks.

Since the propagation mechanisms of deflagrations and detonations
differ so greatly, there is no reason to believe that flammability limits
and detonability limits must coincide. In general, it has been found
that detonability limits are somewhat narrower than flammability limits,
in that there are mixtures that can be made to deflagrate but not to
detonate. The reason for this is unclear; in principle it would seem that
detonable mixtures may exist that cannot support deflagrations. More
research on these limit questions is needed.

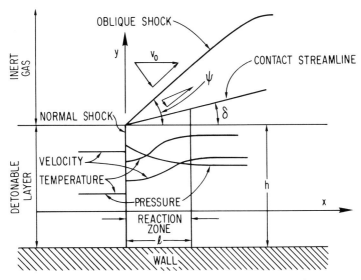

Fig. 4.10. Schematic diagram of a model for a detonation adjacent to a compressible medium.

The boundary layer at the wall of a tube decreases the velocity of propagation of a detonation through the tube, below the Chapman-Jouguet value, by an amount that varies inversely with the tube diameter and for which a reasonable degree of understanding is available (Williams, 1985). Such decreases also can be associated with incomplete combustion; there have been observations of unusually low-velocity detonations that must involve incomplete combustion but that are not well understood. If the tube diameter is made too small, then a detonation will fail to propagate through it. Similarly, a cylinder of solid or liquid explosive with a diameter below the failure diameter cannot be detonated. Failure occurs more readily in the absence of confinement because cooling is associated with lateral gas expansion behind the leading shock, as illustrated in Fig. 4.10.

An understanding of the failure mechanism can be obtained by the same kinds of ideas that explain the reduced velocity for detonations propagating down narrow tubes. Following the leading shock, the heat release is considered to occur in a variable-area, quasi-one-dimensional, inviscid flow. This heat release must increase the Mach number to unity to maintain a Chapman-Jouguet wave. For constant-area, subsonic flow with heat addition, it is known that the heat release increases the Mach number (along the Rayleigh line). For variable-area, isentropic, subsonic flow, an area increase is known to decrease the Mach number. The contact streamline shown in Fig. 4.10 illustrates that for a combustible layer of thickness h adjacent to a nonreactive gas, the oblique shock in the nonreactive gas induces an area increase of the flow in the reaction zone. If in the reaction zone the effect of the area increase on the Mach number exceeds that of the heat release, then the Mach number will decrease, and the wave must fail. An approximate criterion for this to occur is

$$\frac{dT^o}{T^o} \lesssim 2\frac{dA}{A} , \tag{4.31}$$

where T^o is the stagnation temperature and A the cross-sectional area. When all gas properties are the same, this criterion is roughly

$$\ell > h\gamma/2 . \tag{4.32}$$

Thus, the total reaction-zone length is compared with the layer thickness to identify failure conditions. The comparison for detonations in tubes is with the boundary-layer thickness, instead, and the analysis is more involved but results in successful propagation under confinement more easily than in the open.

The failure conditions just given are rough estimates. More precise analyses show that the critical conditions depend strongly on the properties of the gas outside the combustible layer. Because of the short passage time of the reaction zone in a detonation, inertial properties of the bounding material, rather than its mechanical strength, determine whether the material provides confinement during detonation passage. If the mass per unit area of a wall is greater than roughly ℓ times the initial density of the combustible, then the detonation will behave as if it were confined. Thus thin plastic sheets, incapable of withstanding detonation pressures or of contributing to detonation development, can cause a gaseous detonation, once established, to propagate as if it were under confinement. These aspects of detonation dynamics need to be taken into consideration in planning for detonation experiments and in safety questions.

Outstanding Problems in Ignition, Explosions, and Detonations

Tables 4.2 and 4.3 provide lists of some current research problems in the areas of ignition, explosions, and detonations. These lists, while not exhaustive, identify representative topics on which more needs to be learned by further study. Many, but not all, of these topics have been discussed in the preceding presentation. Additional information about these problems is available in the more recent bibliographical citations that follow the tables.

Table 4.2. Current problems in ignition and in explosions that deserve further attention

1. Heat loss by forced or natural convection during ignition
2. Ignition development at hot isothermal boundaries
3. Non-one-step chemical-kinetic effects in ignition
4. Effects of size and shape of hot body or of reactive medium in ignition
5. Non-one-dimensional motions of reactive gases during ignition; forced or natural convection
6. Pressure-wave generation and propagation during ignition in symmetric configurations
7. Pressure-wave generation and propagation during ignition configurations without symmetry

Table 4.3. Current problems in detonations that merit further consideration

1. Approximations for structure of steady, planar detonation
2. One-dimensional stability of planar detonation
3. Three-dimensional stability of planar detonation
4. Direct initiation of stable detonation
5. Direct initiation of unstable detonation
6. Transmission of detonation from a tube into the open
7. Quenching of detonation by inert gas

Bibliography

Berthelot, M. and Vieille, P. 1881. *Compt. Rend. Acad. Sci., Paris* **93**, 18.

Chapman, D.L. 1899. *Phil. Mag.* **47**, 90.

Choi, Y.S. and Majda, A.J. 1989. *SIAM Review* **31**, 401.

Dold, J.W. 1989. *SIAM J. Appl. Math.* **49**, 459.

Döring, W. 1943. *Ann. Physik* **43**, 421.

Frank-Kamenetskii, D.A. 1969. *Diffusion and heat transfer in chemical kinetics.* Plenum Press, New York.

Gray, P. and Lee, P.R. 1967. *Oxid. Combust. Rev.* **2**, 1.

Jackson, T.L., Kapila, A.K., and Stewart, D.S. 1989. *SIAM J. Appl. Math.* **49**, 432.

Jouguet, E. 1905. *J. Mathématique* **6**, 347.

Kassoy, D.R. and Liñán, A. 1978. *Quart. J. Mech. Appl. Math.* **31**, 99.

Kassoy, D.R. 1985. *A. Rev. Fluid Mech.* **17**, 267.

Kassoy, D.R., Kapila, A.K., and Stewart, D.S. 1989. *Combust. Sci. Technol.* **63**, 33.

Lee, J.H. 1984. *A. Rev. Fluid Mech.* **16**, 311.

Lewis, B. and von Elbe, G. 1987. *Combustion, flames and explosion of gases*, 3rd ed. Academic Press, New York.

Liñán, A. and Williams, F.A. 1971. *Combust. Sci. Technol.* **3**, 91.

Mallard, E. and le Chatelier, H.L. 1881. *Compt. Rend. Acad. Sci., Paris* **93**, 145.

Semenov, N.N. 1935. *Chemical kinetics and chain reactions.* Oxford University Press, London.

Strehlow, R.A. 1984. *Combustion fundamentals.* McGraw-Hill, New York.

Vincenti, W.G. and Kruger, Jr., C.H. 1965. *Introduction to physical gas dynamics.* Wiley, New York.

von Neumann, J. 1942. Theory of detonation waves. Prog. Rept. No. 238; O.S.R.D. Rept. No. 549, Ballistic Research Laboratory File No. X-122, Aberdeen Proving Ground, MD.

Williams, F.A. 1985. *Combustion theory,* 2nd ed. Addison-Wesley Publishing Company, Menlo Park, California.

Williams, F.A. 1987. Combustion. *Encyclopedia of physical science and technology,* vol. 3. Academic Press, New York.

Zel'dovich, Y.B. 1940. *Zhur. Eksp. Teor. Fiz.* **10**, 542.

Zel'dovich, Y.B. and Kompaneets, A.S. 1960. *Theory of detonation.* Academic Press, New York.

5

TURBULENT COMBUSTION

In most practical combustion devices the flow is turbulent. Therefore there has always been strong applied interest in turbulent combustion. Thus far we have studiously avoided mentioning turbulence, except where absolutely necessary, because of the complications that it introduces. Important combustion principles can be defined and understood much more easily without considering turbulence. The presentation of these principles has now been completed, and the stage therefore is set to proceed to address questions of turbulent combustion. This subject is approached here keeping in mind the basic combustion ideas of the previous chapters. The orientation will stress fundamentals and the inherent scientific interest in this complex subject, paying less attention to seeking answers to practical problems.

Turbulence itself is a challenging subject with many different avenues toward partial understanding. Addition of combustion to turbulence increases the complexity and opens new avenues of investigation. Knowledge of different possible avenues of approach can enable appropriate selections to be made of methods for seeking deeper understanding or for obtaining practical predictions of turbulent reacting flows. Therefore, after categorizing different types of turbulent combustion and indicating some characteristics of turbulent combustion processes, we shall offer a classification of approaches to the theory. Future study through these and perhaps other kinds of approaches may be expected to reveal unanticipated aspects of turbulent combustion. The books of Libby and Williams (1980) and of Kuznetsov and Sabelnikov (1986) are devoted to the subject, and a new book (Libby and Williams, 1993) provides updated information.

Regimes of Turbulent Combustion

It is of greatest importance to recognize that there are different regimes of turbulent combustion because the appropriate methods of analysis and ways of thinking about the subject differ for the different regimes. In recent years several different approaches to the classification of regimes have been proposed (Ballal and Lefebvre, 1975; Libby and Williams, 1980; Williams, 1985A,B; Peters, 1987). Only two of the most general approaches will be discussed here. One adopts a practical viewpoint, most directly connected to applications, and the other is based on a fundamental viewpoint that expresses conditions in nondimensional variables. The classification from the viewpoint of practice will be discussed first.

To characterize turbulence it is necessary to give not only the magnitudes of the velocity fluctuations but also their length scales. Therefore, at least two quantities, a velocity and a length, must be specified to determine the turbulence. This motivates considering a plane of the characteristic size L of the overall extent of the combustion process (for example, a characteristic length of a combustion chamber) and the characteristic average flow velocity U in the combustion process (for example, the average volume flow rate divided by the cross-sectional area of a combustion chamber). Figure 5.1 places various practical turbulent-combustion systems in this plane (Goulard, Mellor, and Bilger, 1976).

It is easy to locate regions of different applications in Fig. 5.1. We need only ask first how large is the device in which we are interested and second how high is the velocity through it. Thus spark-ignition engines

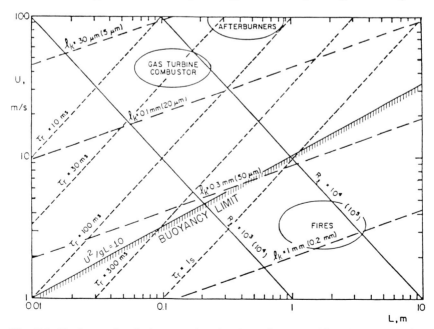

Fig. 5.1. Regimes of turbulent combustion in a diagram of length and velocity.

are seen to fall at the smaller sizes and intermediate velocities, while gas turbines and afterburners lie at somewhat larger sizes and much higher velocities; the sizes of most hostile fires of practical concern are even larger, but the associated velocities are lower. Although placement of the system in the diagram is straightforward, identification of the type of turbulent combustion process requires more consideration.

A significant question concerns the role of buoyancy in the combustion process. The importance of buoyancy is measured by the Froude number F, defined in Chapter 1. A Froude number based on the coordinates of Fig. 5.1 is U^2/gL, where g is the earth's acceleration of gravity (9.8 m/s). A line along which this Froude number equals 10 is drawn in Fig. 5.1 and identified as the buoyancy limit. Below this line effects of buoyance amount to 10% or more and cannot be neglected. Thus it is seen that although buoyancy is unimportant in most engine applications, it cannot be neglected in fires, where it often is the main driver for generating the velocity U.

Lines of constant values of various other parameters are drawn in Fig. 5.1. The residence time τ_r is L/U, a measure of the total time available for the combustion process to occur. These times are seen to range typically from a few milliseconds to occasionally slightly more than one second. Also shown are lines of constant values of the Reynolds number R_ℓ based on the fluctuation velocity and the integral scale ℓ of turbulence in the combustion region. The latter is defined in terms of an integral of a two-point average of fluctuation velocities and typically is somewhat less than L, although of roughly the same order of magnitude. The fluctuation velocity, typically about $0.1U$, can be estimated as $\sqrt{2k}$, where the turbulent kinetic energy is

$$k = \frac{1}{2}\overline{(v - \overline{v})^2} \, , \qquad (5.1)$$

with the bar signifying an average. The Reynolds number R_ℓ then is

$$R_\ell = \sqrt{2k}\,\ell/\nu \, , \qquad (5.2)$$

where ν is the average kinematic viscosity of the gas (roughly 1 cm^2/s in typical hot flows at atmospheric pressure). The Reynolds number R_ℓ is more closely related to the turbulent combustion processes occurring locally than is the engineering Reynolds number UL/ν, based on the macroscopic parameters, and therefore R_ℓ is a parameter of more fundamental significance. Two values of R_ℓ are given, one (not in parentheses) corresponding to atmospheric pressure and the other (in parentheses) corresponding to 10 atm. The inverse dependence of ν on p produces this difference; the range of 1 to 10 atm is roughly representative of the

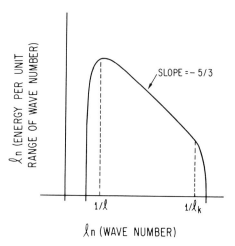

Fig. 5.2. Schematic illustration of the spectrum of turbulent kinetic energy.

range encountered in most applications. It is seen that R_ℓ tends to be large in applications. For turbulence to occur at all we need $R_\ell \gtrsim 1$, since otherwise fluctuations are rapidly damped by viscosity.

For $R_\ell \gg 1$ turbulence is known to exhibit a range of eddy scales. The integral scale ℓ is representative of the size of the largest eddies, and the Kolmogorov scale ℓ_k is of the order of the size of the smallest eddies. The dependence of the turbulent kinetic energy on the scale can be represented in a Fourier-transform space, in which turbulent kinetic energy per unit range of wave number is plotted as a function of the wave number, as shown in Fig. 5.2 in a log-log graph. Dimensional analysis led Kolmogorov to predict the $-5/3$ slope shown in Fig. 5.2 for the inertial (or Kolmogorov) subrange in which the dependence of the energy on the wave number is affected only by the inertial terms in (1.2). The nonlinearities of the Navier-Stokes equations are viewed as causing a cascade of sizes in this subrange, with only the average rate of dissipation, $(2k)^{3/2}/\ell$, being relevant. The viscosity becomes important for dissipation only in the smaller eddies, of size ℓ_k, the length $[\nu^3\ell/(2k)^{3/2}]^{1/4}$ formed from the kinematic viscosity and the dissipation rate; and the dissipation is so efficient there that there is a sharp, exponential, cutoff of energy, giving negligible fluctuations at sizes smaller than ℓ_k. The behavior in Fig. 5.2 has been established experimentally only for nonreacting, constant-density turbulence with negligible buoyant stratification, and therefore there is some uncertainty about its application in combustion. Nevertheless, it seems likely to remain roughly applicable, so that the cutoff at ℓ_k remains. In Fig. 5.1 lines of constant values of ℓ_k are shown for 1 and 10 atm to provide an indication of the smallest sizes that need to be considered (from the viewpoint of turbulence).

Figure 5.1 involves dimensional quantities. A more fundamental classification can be presented in a dimensionless plot. A nondimensional parameter characteristic of combustion is a Damköhler number, the ratio of a flow time to a chemical time. The chemical time will be denoted by τ_c, and although there are many such times, associated with different reactions, species concentrations, and temperatures, only one is introduced. This chemical time is to be considered to be representative of the overall chemical process and is to be defined more precisely in specific examples. The flow time may be taken as the turnover time of the largest eddies, $\ell/\sqrt{2k}$, to give the Damköhler number based on the integral scale of the turbulence as

$$D_\ell = (\ell/\sqrt{2k})/\tau_c . \qquad (5.3)$$

Figure 5.3 is a plane of R_ℓ and D_ℓ in which turbulent combustion systems can be placed. The character of the turbulent combustion is quite different in different parts of this plane (Williams, 1985B).

The principal difference in combustion regimes is that between large and small Damköhler numbers. For sufficiently small Damköhler numbers all of the fluid dynamics is rapid compared with all of the chemistry, and so the turbulent mixing smooths out the concentration fluctuations before the combustion occurs. The chemistry then occurs slowly and in a distributed manner, throughout the region over which the turbulence has carried the reactants. This therefore is a regime of distributed reactions, in which the chemistry proceeds slowly over large regions of the flow.

In the opposite limit of sufficiently large Damköhler numbers, all of the chemistry is rapid compared with all of the turbulence. This has different consequences in detail for premixed flames and diffusion flames, for example because the former cannot have fast chemistry everywhere. However, in general it implies that the reactions occur in thin sheets moved around by the turbulence. For diffusion flames these sheets are where the fuel and oxidizer diffuse into each other, while for premixed flames the sheets are where wrinkled laminar flames are propagating. Thus there is a regime of reaction sheets in which the fast chemistry occurs only in the wrinkled sheets and not elsewhere.

Although the distributed-reaction and reaction-sheet limits clearly exist, it is difficult to know just how small or large D_ℓ must be to be sure of achieving them. A parameter other than D_ℓ may be more relevant. Instead of a turnover time for the largest eddies, a characteristic time for the smallest eddies could be more important. This shorter time is

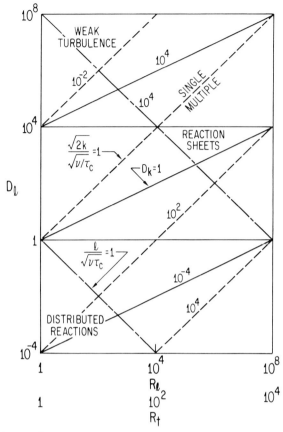

Fig. 5.3. Regimes of turbulent combustion in a diagram of Reynolds number and Damköhler number.

the Kolmogorov time $\sqrt{\nu\ell/(2k)^{3/2}}$, the shortest time scale of the turbulence. The Damköhler number based on the Kolmogorov time is

$$D_k = \sqrt{\nu\ell/(2k)^{3/2}}/\tau_c = D_\ell/\sqrt{R_\ell}\,, \qquad (5.4)$$

the last equality of which has been used to draw lines of constant values of D_k in Fig. 5.3. Since the Kolmogorov time is the shortest, large values of D_k are sufficient to assure that the chemical time is small compared with all turbulence times. Therefore $D_k \gg 1$ appears to be a sufficient condition for occurrence of the reaction-sheet regime.

Conversely, $D_\ell \ll 1$ may be thought to be a sufficient condition for the occurrence of the distributed-reaction regime. However, Damköhler (1940) in his original considerations of premixed turbulent combustion suggested that the integral scale ℓ should be compared with the thickness of a premixed laminar flame δ, which was found in Chapter 2 to be roughly of the order of $\sqrt{\nu \tau_c}$. From the definitions of D_ℓ and R_ℓ it is seen that

$$\sqrt{R_\ell D_\ell} = \ell/\delta \tag{5.5}$$

in these rough approximations, so that Damköhler's criterion for the occurrence of the distributed-reaction regime in premixed turbulent flames, $\ell/\sqrt{\nu \tau_c} \ll 1$, is more stringent than $D_\ell \ll 1$. This is also shown by the lines of constant values of $\ell/\sqrt{\nu \tau_c}$ plotted in Fig. 5.3, as calculated from (5.5). It seems safe to say that $\sqrt{R_\ell D_\ell} \ll 1$ always will be a sufficient criterion for occurrence of the distributed-reaction regime, and in fact it will be seen in the following section that $D_\ell \ll 1$ probably is sufficient.

For premixed flames D_k has an alternative interpretation as $(\ell_k/\delta)^2$, since

$$\ell_k = \ell/R_\ell^{3/4} \tag{5.6}$$

by its definition. Thus for $D_k > 1$ all turbulence scales are larger than the premixed laminar flame thickness, and the turbulence is expected to wrinkle the laminar flames without breaking them.

Another parameter of relevance to premixed turbulent combustion is $\sqrt{2k}/\sqrt{\nu/\tau_c}$, the ratio of an average fluctuation velocity of the turbulence to the laminar burning velocity. Constant values of this parameter also are plotted in Fig. 5.3. When this parameter is very small, laminar flames can propagate through the turbulence with little change in their velocity; this limit of the reaction-sheet regime is identified as that of weak turbulence in Fig. 5.3. Setting this parameter equal to unity defines roughly a boundary between single and multiple reaction sheets for premixed flames, as indicated in Fig. 5.3. The basis for this division is that when the parameter is large the turbulence can move the wrinkled laminar flames large distances and cause them to bend back on themselves, possibly cutting off pockets of unburnt gas (but *never* of burnt gas, so long as the sheets remain continuous and propagate into the unburnt mixture). These distinctions do not apply to diffusion flames, because laminar diffusion flames are not propagating waves.

Some investigators prefer to employ the Taylor scale ℓ_t of turbulence instead of ℓ. In deriving an equation for the average rate of dissipation of turbulent kinetic energy, the average of (1.2) is subtracted from (1.2) to obtain a differential equation for the evolution of the velocity fluctuations, and this difference is dotted into $(v - \overline{v})$ then averaged. The resulting equation contains the dissipation term $\overline{\nu |\nabla \cdot (v - \overline{v})|^2}$, which is approximated as $\nu(2k)/\ell_t^2$ to define the Taylor scale. Equating this average dissipation rate to $(2k)^{3/2}/\ell$, we find that

$$\ell_t = \ell/\sqrt{R_\ell} , \tag{5.7}$$

which shows by comparison with (5.6) that the Taylor scale is an average turbulence length that falls between ℓ_k and ℓ. The Reynolds number based on the Taylor scale is defined as

$$R_t = \sqrt{2k}\,\ell_t/\nu \qquad (5.8)$$

and is seen from (5.2) and (5.7) to be $\sqrt{R_\ell}$, so that the horizontal scale in Fig. 5.3 can be interpreted in terms of R_t, as indicated.

The question arises as to whether reaction-sheet or distributed-reaction regimes are most likely to be encountered in practice. Although some clarification of this question has been obtained recently, much uncertainty remains. Table 5.1 summarizes premixed-flame and diffusion-flame applications, placing them in the reaction-sheet or distributed-reaction regime, or both. An entry under both headings means that either there are applications in both regimes, or else the proper regime just is not known (usually the latter). Well-stirred chemical reactors, widely employed in the chemical processing industry (for example, jet-stirred reactors, in which many small, high-speed jets of reactants enter an enclosure at various angles to promote strong mixing) are designed to operate in the distributed-reaction regime and usually do so. Recent detailed studies of spark-ignition engines have shown that in none of these applications can the combustion be said to definitely occur in the distributed-reaction regime, but in many (probably most) of them, the combustion definitely occurs in the reaction-sheet regime; some of the smaller, high-speed engines (such as motorcycle engines or especially model airplane engines) are estimated to lie in the intermediate range where the regime is uncertain. In large fires, large furnaces, and marine Diesels, the flow

Table 5.1. Regimes encountered in applications of turbulent combustion

	Reaction Sheet	Distributed Reaction
Premixed	Spark-ignition engines	Well-stirred reactor
	Turbojets	Turbojets
	Ramjets	Ramjets
	Afterburners	Afterburners
	Rockets	Rockets
Diffusion	Fires	Well-stirred reactor
	Furnaces	
	Marine diesels	
	Jet flames (flares, oil wells)	
	Chemical lasers	Chemical lasers
	Rockets	Rockets
	Supersonic combustion	

times are long enough that reaction-sheet combustion is expected to develop for the principal heat-release chemistry; on the other hand, there
is fuel-rich chemistry in these applications, leading to soot and (at least
in the engines) to preignition events, chemistry that has longer chemical
time scales and does not occur in a reaction-sheet mode. Thus, from
the applications about which most is known, it is seen that some systems can be identified that definitely exhibit reaction-sheet combustion
and others that definitely experience distributed-reaction combustion.
Therefore there is practical interest in both limiting regimes.

In the region of Fig. 5.3 between the line $D_k = 1$ and the line
$\ell/\sqrt{\nu\tau_c} = 1$ is a range, very broad at high R_ℓ, in which the regime of
turbulent combustion is uncertain. A few ideas are becoming available
about possible characteristics of premixed turbulent combustion in this
range (Peters, 1987), as will be mentioned later. Because of the differences between premixed flames and diffusion flames, it is best to focus
separately on each of these categories in the following discussions.

Turbulent Premixed Flames

Figure 5.4 serves to emphasize differences between premixed turbulent combustion in different regimes. It shows results of experimental
measurements of turbulent burning velocities v_T under a wide range of
conditions, performed by Ballal and Lefebvre (1975). The assumption
generally has been made that in turbulent flows a burning velocity exists,
just as it does in laminar flows, giving the average rate of propagation
of the flame through the turbulent premixed gas. In the reaction-sheet
regime, this assumption is supported although not proven by the observation that the wrinkled laminar flames possess a burning velocity.
In the distributed-reaction regime, it is less certain how good the assumption may be, because the unique structural aspects of the laminar
flame that lead to the existence of the burning velocity v_o no longer
are present; in particular, the apparently essential preheat zone may be
absent. Nevertheless, turbulent-flame propagation experiments typically
have fixed propagation distances, and flames in steady turbulent flows
will spread at a measurable average angle across a duct, so that experimental values for v_T can be obtained even if it does not exist in the
same fundamental sense as v_o.

The results for v_T in Fig. 5.4 exhibit three distinct types of behavior. The points for "low-intensity turbulence" have intensities too high
to fall in the weak-turbulence limit and scales small enough for l_k to
be comparable with δ, so that the reaction-sheet regime may not apply,
or if it does, influences of hydrodynamic instability may be especially

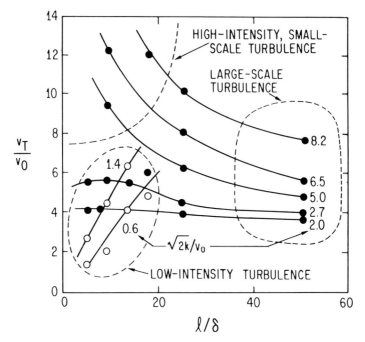

Fig. 5.4. Measured turbulent burning velocities as functions of turbulence scale for various turbulence intensities.

significant; a strong increase in v_T/v_o with increasing scale is observed for these points. On the other hand, for "high-intensity, small-scale turbulence," v_T/v_o was observed to decrease with increasing scale; this may reflect distributed-reaction behavior, while the former (opposite) behavior may correspond to an intermediate regime. The reaction-sheet regime is estimated to occur at the largest scales, and from the points for "large-scale turbulence" in Fig. 5.4 it is seen that v_T/v_o tends to become independent of the scale in this limit, varying only with intensity. It should be emphasized, however, that because of the higher R_ℓ for these points, none of them are estimated to have $\delta < \ell_k$; only the points for $\sqrt{2k}/v_o = 0.6$ are estimated to obey this inequality, and all of the points have δ within about a factor of two of ℓ_k. Coupled with inaccuracies associated with experimental difficulties, these observations imply that only general trends should be sought from Fig. 5.4. Nevertheless, for the "large-scale turbulence" part of Fig. 5.4, most eddies will be large compared with δ, and the experiments suggest a scale-independent increase in v_T/v_o with increasing turbulence intensity. Theories of premixed turbulent flame propagation will be seen to predict an increase in v_T/v_o with increasing intensity $\sqrt{2k}/v_o$ in this regime, qualitatively in accord with experiment. Theories for the other two types of behavior

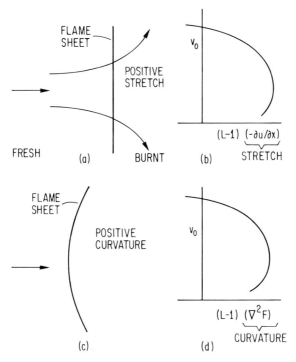

Fig. 5.5. Influences of flame stretch and of flame curvature on premixed laminar flamelets.

seen in Fig. 5.4 are poorly developed, despite fairly extensive literature espousing clever and carefully developed ideas that rest on questionable foundations.

As D_k is decreased, departures from the reaction-sheet regime in Fig. 5.3 occur by punching holes in the reaction sheets. These holes develop through laminar-flame extinctions produced by the stretch and curvature of the flame, introduced by the turbulence, as illustrated in Fig. 5.5. Diverging streamlines, as in a stagnation-point flow, represent a strain field that is said to "stretch" the flame, as illustrated in Fig. 5.5a, by tending to move adjacent fluid elements on the reaction sheet away from each other. This corresponds to positive stretch; converging streamlines introduce negative stretch, sometimes called compression. Positive stretch means positive values for the two-dimensional divergence in planes parallel to the flame sheet, $\nabla_\perp \cdot v_\perp$ (when the subscript $\perp$ identifies these two-dimensional components, perpendicular to the normal to the flame sheet). Through mass conservation (1.1) this implies a negative value for the normal component, $\partial u / \partial x$, so that the flame-sheet thickness tends to be decreased. Associated with this decrease are increased gradients within the flame, and therefore increased rates of molecular transport. For Lewis numbers greater than unity this gives more conductive heat loss than heat generation by reactant diffusion and a consequent decrease in the flame temperature T_f, as explained in Chapter 2. Therefore the burning velocity is decreased by positive stretch if $L > 1$, approaching an extinction point, as illustrated in Fig. 5.5b (Sivashinsky, 1983; Clavin, 1985; Williams, 1985A).

A similar effect results from flame curvature. If the wrinkled flame-sheet shape is described locally by setting x equal to a function $F(y, z, t)$ with $x > F$ in the burnt gas, then the curvature toward the burnt gas is $\nabla_\perp^2 F$ and is positive for the shape illustrated in Fig. 5.5c. As reasoned in Chapter 2, for $L > 1$ this curvature reduces v_o and leads toward the extinction illustrated in Fig. 5.5d. A nondimensional measure of the combined effects of stretch and curvature, sometimes called the Karlovitz number after the investigator who first introduced the term "flame stretch" (Karlovitz et al., 1953), is

$$\kappa = (\delta/v_o)(v_o \nabla \cdot \boldsymbol{n} - \boldsymbol{n} \cdot \boldsymbol{\Phi} \cdot \boldsymbol{n}) \,, \tag{5.9}$$

where the strain-rate tensor appearing in (1.6) is

$$\boldsymbol{\Phi} = \tfrac{1}{2} \left[(\nabla \boldsymbol{v}) + (\nabla \boldsymbol{v})^T \right] \,, \tag{5.10}$$

and $\boldsymbol{n}$ is the unit vector normal to the flame sheet, pointing toward the fresh mixture and given by $\nabla_\perp F$. For small values of κ, called weak stretch, the effects of both stretch and curvature on the flame depend only on κ, but for moderate stretch (κ of order unity) or strong stretch (κ large), stretch and curvature exert quantitatively different influences.

Analyses of laminar-flame structures by AEA (see Chapter 2) have been completed for weak, moderate, and strong stretch. These analyses have included effects of $L \neq 1$, effects of general strain-rate tensors, and of swirl having vorticity vectors normal to the flame (Libby, Williams, and Sivashinsky, 1990), and effects of nonadiabaticity, that is, product-gas temperatures T_∞ different from the adiabatic flame temperature T_{af}. Representative results for $L = 1$ are shown in Fig. 5.6, which demonstrates abrupt extinction for sufficiently subadiabatic conditions (Williams, 1985A). For adiabatic flames, AEA shows abrupt extinctions not to be achievable for $L < 1$ and to be achievable with difficulty for $L > 1$. At large κ gradual extinction is found to occur for all values of H and L. Experimentally, stretched laminar flames are observed to experience abrupt extinction. This may be due partially to subadiabaticity through radiant energy loss, but also to enhanced diffusive loss of active intermediates from the reaction zone by the higher gradients. Analyses of stretched flame structures with detailed chemistry are needed to better evaluate the roles of the intermediates. The character of the strain

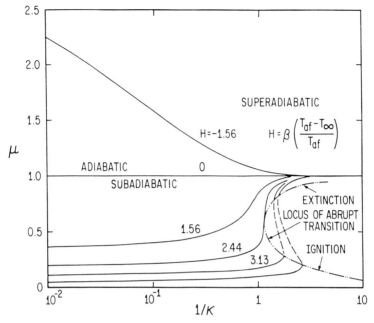

Fig. 5.6. The ratio μ of the rate of heat release per unit area of a strained premixed flamelet to that of an unstrained flame as a function of Damköhler number for various nonadiabaticity parameters (β = Zel'dovich number of Chapter 2).

fields in turbulent flows is a topic of continuing investigation that is relevant to the influences of the turbulence on wrinkled flames. A single parameter, usually called c, measures the ratio of strain rates in two principal orthogonal directions parallel to the plane of the stretched laminar flame. The range of c is $-1 < c < 1$, with $c = 1$ corresponding to axisymmetric stagnation flow, $c = 0$ to planar stagnation flow, and $c < 0$ corresponding to inflow in one of the two directions along the flame. For $c = -1/2$ the strain rate for the inflow equals the component of the strain rate normal to the flame, and for $c = -1$ the tangential inflow and outflow strain-rate components are equal in magnitude. A planar flame is known to align rapidly normal to the largest negative principal rate of strain, and therefore only the range $-1/2 \lesssim c \lesssim 1$ is relevant to plane-flame behavior. For $c = -1/2$, the two equal inflow strain rates are expected to lead to cylindrical flames which, like back-to-back planar flames, are always predicted to exhibit abrupt extinction at sufficiently high strain rates. For $-1/2 < c < 0$ the flames might be planar or elliptic-like cylinders, while for $0 \lesssim c \lesssim 1$ generally planar stretched flames are anticipated. There are distributions of c in turbulent flows, and experiments at high Reynolds numbers have suggested a most probable value of $c = -1/2$, corresponding to stretched, symmetrical vortices (called "spaghetti"). Cylindrical flames wrapped around vortex cores are expected in this turbulence. Numerical simulation through computer solution of the Navier-Stokes equations (Ashurst et al., 1987) has suggested a most-probable value of $c \approx 0.3$, nearly flat but with two different outflow components (called "linguini"). Flat flames are expected in turbulence of high Reynolds number of this type. More research is needed in characterizing turbulence strain fields before definitive statements can be made about expected wrinkled-flame shapes.

Vortices can affect structures of wrinkled flames in turbulent flows. Flat flames can be spun into unstretched vortex cores, eventually giving a core of burnt gas with a flame propagating outward from it (Peters and Williams, 1989). Stretch of the vortex can arrest the outward propagation. Also, the radial pressure field of a vortex depends on the gas density, and it can be reasoned that if part of the vortex core is fresh mixture and another part burnt gas, then the products will be driven by this pressure field along the core into the unburnt mixture at velocities that can greatly exceed the laminar burning velocity (Chomiak, 1990). These aspects of influences of vortices in turbulence on premixed flames are in early stages of investigation. In addition to theory, experiments on flames in individual vortices are being pursued. There certainly is more to be learned about the interactions of the rotational fields in turbulence with flames.

Some new ideas recently have been put forward (Peters, 1987) concerning two regions in Fig. 5.3. One is the reaction-sheet region of multiple sheets, between $D_k = 1$ and $\sqrt{2k}/\sqrt{\nu/\tau_c} = 1$. For $D_k < 1$, holes prevent the sheets from being described as continuous surfaces. For $\sqrt{2k}/\sqrt{\nu/\tau_c} < 1$, there is a single, well-defined sheet with a wrinkled shape influenced appreciably by the laminar flame propagation. In between, the sheets are continuous but highly irregular (and possibly not simply connected) because of the strong effect of the turbulence on their motion. In this region, it is attractive to try to describe the flame shape as a fractal. Ideas of fractals (objects with noninteger space dimensions) have arisen in turbulence in various ways, and the flame sheet in this regime of turbulent combustion is a most natural application (Kerstein, 1988B, 1991A; Mantzaras, Felton, and Bracco, 1989; Murayama and Takeno, 1989). However, a fractal description of the sheet can apply at best only over a limited range of scales. There is a maximum size, the integral scale ℓ, beyond which there are no turbulent fluctuations, hence no fractal description. There is also a minimum size, recently called the Gibson scale, below which the laminar flame velocity exceeds the eddy velocity, preventing the development of a fractal by the random turbulent displacements. Since the eddy velocity is the cube root of the product of the eddy size and the average rate of dissipation of turbulent kinetic energy, the Gibson scale is

$$\ell_g = v_o^3 / \left[(2k)^{3/2} / \ell \right] = \ell_k D_k^{3/2} , \qquad (5.11)$$

where the approximation $v_o = \sqrt{\nu/\tau_c}$ has been employed along with (5.2), (5.4), and (5.6). Fractal descriptions can work only for scales between ℓ_g and ℓ, which define the fractal cutoffs. This range disappears at the boundary $\sqrt{2k}/v_o = 1$ of single sheets, where ℓ_g reaches ℓ. At the opposite boundary, $D_k = 1$, ℓ_g reaches ℓ_k, and beyond this ($\ell_g < \ell_k$) ℓ_g loses significance because it is smaller than the smallest turbulence scales. If sheets with holes are described as fractals, then the description could hold over the entire turbulence range of scales, from ℓ_k to ℓ, were it not for the holes deleting the smaller scales, as reasoned below. Kuznetsov and Sabelnikov (1986) identify a scale closely related to ℓ_g but having a different physical interpretation, involving hydrodynamic instability.

The other region in Fig. 5.3 is that between $D_\ell = 1$ and $D_k = 1$. In this region the smaller eddies poke holes in the sheet, but sheets are still possible in the larger eddies, which have smaller strain rates. The characteristic time of an eddy is the cube root of the ratio of the square of the size to the average rate of dissipation of turbulent kinetic energy, and a rough criterion for breaking the flame is to equate the reciprocal of this time (as an eddy strain rate) to the reciprocal of the chemical time τ_c, giving what has recently been called the Peters scale,

$$\ell_p = \sqrt{\tau_c^3 [(2k)^{3/2} / \ell]} = \ell_k / D_k^{3/2} . \qquad (5.12)$$

There can be significant numerical differences between τ_c^{-1} and the extinction strain rate, just as between v_o and $\sqrt{\nu/\tau_c}$ (see Description of Flame Structure in Chapter 2), but for a simplified systematization it seems best to overlook these differences here with the realization that limits can be wrong numerically by an amount on the order of a factor of ten. The result is that, between $D_\ell = 1$ and $D_k = 1$, reaction sheets may still be anticipated in eddies of size between ℓ_p and ℓ, but not in eddies of size between ℓ_k and ℓ_p. In these smaller eddies, fuel diffuses into the hot products without burning, and combustion in a distributed-reaction type of mode might then occur later in these quenched regions. The size range for a fractal description would be between ℓ_p and ℓ, since the character of the structure clearly must change at the size ℓ_p. At the boundary

$D_k = 1$, where the holes first begin to appear, ℓ_p equals ℓ_k, and the sheets become possible in all eddies. At the opposite boundary $D_\ell = 1$, from (5.3) and (5.12) ℓ_p becomes equal to ℓ, and beyond this ($\ell_p > \ell$) reaction sheets cannot exist in any eddies. Because the sheets are sensitive to stretch, this boundary differs from the Damköhler boundary $\ell = \delta$, and there is a region in Fig. 5.3 between $D_\ell = 1$ and $\ell = \delta$ in which no wrinkled flames are present, even though eddies larger than a premixed laminar flame thickness are present.

Although ideas are rapidly emerging for describing turbulent flame propagation in the region between $D_\ell = 1$ and $D_k = 1$ (Peters, 1987), these ideas are insufficiently developed to be discussed here. Therefore attention from now on in this section will be restricted to the reaction-sheet regime, $D_k > 1$. In this regime, a field equation is readily written for describing the dynamics of the wrinkled flame in the turbulent flow (Markstein, 1964; Williams, 1985B; Kerstein, Ashurst, and Williams, 1988). A continuous and differentiable function $G(x, t)$ can be introduced such that G maintains a constant value at the reaction sheet, say $G = 0$, and $G < 0$ in the fresh mixture, while $G > 0$ in the burnt gas. Then with the evolution equation

$$\partial G/\partial t + v \cdot \nabla G = V|\nabla G| , \qquad (5.13)$$

where V is the local propagation velocity of the laminar flame, the function G will maintain its constant value on the reaction sheet. Effects of strain and curvature on the laminar flame propagation can be included through a perturbation approach by putting

$$V = v_o(1 - b\kappa + \cdots) \approx v_o e^{-b\kappa} , \qquad (5.14)$$

where v_o applies to the steady, planar flame, and κ is given by (5.9), with the constant b (the Markstein number) determined by analysis of the influence of stretch on the burning velocity. Analysis by AEA for small κ has given

$$b = \left(\frac{T_\infty}{T_\infty - T_o}\right) \int_{T_0}^{T_\infty} \left(\frac{\lambda}{\lambda_o}\right) \frac{dT}{T}$$

$$+ \frac{\beta}{2}(L-1)\left(\frac{T_o}{T_\infty - T_o}\right) \int_{T_0}^{T_\infty} \left(\frac{\lambda}{\lambda_o}\right) \ln\left(\frac{T_\infty - T_o}{T - T_o}\right) \frac{dT}{T} , \qquad (5.15)$$

where λ is the thermal conductivity, β the Zel'dovich number of (2.8), and the temperature integrals are to be performed over the preheat zone of the steady, planar, laminar flame. Equation (5.13) along with the nonreacting flow equations (1.1) and (1.2) for the constant-density flows on each side of the wrinkled flame provides a closed set of conservation equations for describing the turbulent flow and the flame motion.

Numerical simulation of turbulent combustion has been undertaken through computer integration of these equations (Kerstein, Ashurst, and Williams, 1988). For simplicity at the beginning, an approximation of constant density was introduced. The turbulence field can then be calculated independently of (5.13), which in effect is superimposed on the turbulence. The results show the shapes of the reaction sheets and can give the turbulent burning velocity. In agreement with Huygens principle, the shapes develop cusps pointing into the burnt gas. More powerful computers could improve the accuracies of these simulations. Methods of the renormalization group also have recently been applied to (5.13) to obtain a formula for the turbulent burning velocity (Yakhot, 1988; Kerstein, 1988A; Sivashinsky, 1988; Majda, 1991). These methods additionally have been introduced in subgrid-scale modeling for large-eddy simulations. Spectral decompositions and moment-method modeling have also been applied recently to (5.13) (Peters, 1993). Definitions of these various methods will be given in the penultimate section of this chapter. Active research is in progress, and additional results may be anticipated in the future.

A simple view of the turbulent burning velocity when $V = v_o$ was recognized by Damköhler (1940) and is illustrated in Fig. 5.7. Since all of the reactant gas passes through the wrinkled flame, two alternative ways of calculating the volume flux give

$$v_T = v_o \overline{(A_f/A)} \,, \tag{5.16}$$

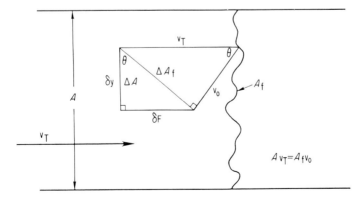

Fig. 5.7. Schematic illustration of wrinkled laminar flame.

the last factor being the average ratio of the wrinkled-flame area to the cross-sectional area. In terms of the function G, for a turbulent flame propagating in a direction defined by the unit vector e, the area ratio is

$$\overline{(A_f/A)} = \overline{(|\nabla G|/|e \cdot \nabla G|)} \, , \tag{5.17}$$

in which the last average is a cross-sectional average taken at values of G corresponding to the reaction sheet. If the sheet does not fold back on itself, then the representation $G = x - F$ can be used, where x is the coordinate in the propagation direction, and F is the function of transverse coordinates introduced before (5.9) to define the wrinkled-flame location. In this case it is found that

$$\overline{(A_f/A)} = \overline{\sqrt{1 + |\nabla_\perp F|^2}} \, , \tag{5.18}$$

as is evident from geometrical reasoning, since $\nabla_\perp F$ defines the local flame tilt. For weak turbulence, in terms of the streamwise Eulerian displacement

$$a = \int (u - \bar{u})dt \, , \tag{5.19}$$

where u is the x component of velocity, it has been shown that $\nabla_\perp F = \nabla_\perp a$, and for small values of this quantity, then

$$\overline{(A_f/A)} = 1 + \tfrac{1}{2}\overline{|\nabla_\perp a|^2} \, . \tag{5.20}$$

With (5.16), these formulas, in the order of decreasing turbulence intensity, give successively simpler expressions for the turbulent burning velocity.

The problem of finding v_T in the reaction-sheet regime has been addressed often by different investigators on the basis of different approximations. As indicated in the discussion of Fig. 5.4, it is expected that in this regime the ratio v_T/v_o is a function only of the ratio $\sqrt{2k}/v_o$, with no dependence on the turbulence scale. If we temporarily call the first of these ratios y and the second x, then the problem appears tantalizingly simple; we need only find the function f in the relationship $y = f(x)$. In fact, the problem is challenging because f depends on properties of both the turbulence and the reaction-sheet evolution. The average $\overline{|\nabla_\perp a|^2}$ in (5.20) can be reasoned from considerations of turbulence to be independent of scale and proportional to x/y; a burning-velocity formula, estimated from (5.16) and (5.18), is then found to be (Clavin, 1985; Williams, 1985A)

$$y = \sqrt{\tfrac{1}{2}\left(1 + \sqrt{1 + 8Cx^2}\right)} \, , \tag{5.21}$$

where C is an undetermined constant that is dependent, for example, on the degree of anisotropy of the turbulence. Equation (5.21) is useful only for small x and is a guess that is likely to be inaccurate for large x. It has recently been challenged even for small x (Kerstein and Ashurst, 1992) on the basis that, to approach stationarity of F, a balance between the first and last terms of (5.13) must develop, giving $y - 1 \sim x^{4/3}$, instead of $y - 1 \sim x^2$, for small x. An empirical formula, often used in applications to spark-ignition engines, is

$$y = 1 + C'x , \qquad (5.22)$$

where C' is a constant. Although the simple linear functional form of (5.22) must be wrong for small x (and possibly also for large x), it still can correlate measurements in spark-ignition engines because the data have large inaccuracies as a result of the severe experimental difficulties. By considering the dynamics of wrinkled flame propagation in turbulence, Klimov (1975, 1983) proposed the approximation

$$y = 2.4x^{0.7} , \qquad (5.23)$$

for large x, which agrees well with an appreciable amount of laboratory data. By analysis through a method for calculating the evolution of probability-density functions (to be discussed in the penultimate section of this chapter), Pope and Anand (1985) found for large x that

$$y = 1.4x , \qquad (5.24)$$

which appears to suffer from a lack of influence of the reaction-sheet evolution. Through the renormalization-group approach, Yakhot (1988) found that

$$y = x/\sqrt{\ln y} , \qquad (5.25)$$

which retains the essential physics and also appears to be in good agreement with those experimental results that most clearly lie in the reaction-sheet regime; however, there are indications (Sivashinsky, 1988; Kerstein and Ashurst, 1992) that assumptions of the analysis need further scrutiny.

Figure 5.8 compares these formulas with recent measurements of $\overline{(A_f/A)}$ in spark-ignition engines. The measurements (Mantzaras, Felton, and Bracco, 1989) are made in an engine with a transparent window by seeding the intake gas with fine particles and recording the scattering of a laser sheet from the particles. The results are seen to agree best with (5.23), although (5.24) and (5.25) also could be acceptable, and (5.22) with an appropriate choice of C' (about 1.4) would correlate the observations. The data suggest a curvature that is not consistent with (5.22)

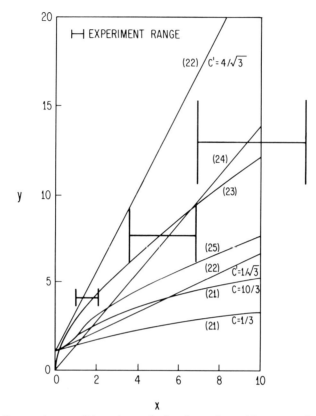

Fig. 5.8. Comparisons of burning-velocity formulas with area-ratio measurements.

and (5.24). It would be helpful to have experimental results with less uncertainty to provide better tests of the formulas, and work is in progress
to obtain such data. Improved experimental results for v_T/v_o directly,
rather than through the area ratio, also would aid in evaluating theories
for the reaction-sheet regime. This might eventually be achievable, even
in cylinders, through laser-doppler velocimetry, combined with newer
laser techniques, currently under development, like those that produced
the data in Fig. 5.8.

    The discussion here has focused largely on the effects of the tur-
bulence on the flame propagation and has bypassed questions of the
effects of the flame on the turbulence. Premixed flames affect turbulence strongly because of the large density decrease and temperature
increase across the flame. Velocity changes across wrinkled flames were
discussed in Chapter 2 in connection with hydrodynamic instability. In
fact, the hydrodynamic instability itself requires thought concerning its
influences on turbulent flame propagation (Sivashinsky, 1983; Clavin,

1985). In the weak-turbulence and single-sheet subregimes of Fig. 5.3, the hydrodynamic instability of planar flames cannot be neglected; either it is stabilized by diffusive-thermal effects at small scales and by favorably oriented buoyancy effects at large scales, or it causes the planar flames to evolve into nonplanar shapes, so that the influences of the relatively weak turbulence on the nonplanar structures need to be investigated. For the multiple-sheet subregime, investigators who have considered the problem have concluded that the hydrodynamic instability is overpowered by the strong turbulence and can be neglected. This conclusion is probably correct, even though a rigorous demonstration of it is unavailable. Sufficiently strong turbulence is estimated to overcome even buoyant instabilities, so that the body force in (1.2) can be neglected in a first approximation.

For a weakly wrinkled flame, the reasoning of Chapter 2 shows that the jump conditions across the flame cause local amplifications of transverse components of velocity fluctuations, with consequent introduction of anisotropy into the turbulence and corresponding increases in the turbulent kinetic energy, sometimes called flame-generated turbulence. There are zones of hydrodynamic adjustment on both sides of the wrinkled flame sheet, and in these zones both the anisotropy of the turbulence and its kinetic energy can change (Aldredge and Williams, 1991). Only a few studies of the character of the hydrodynamic adjustment have been completed; there is more to be learned.

Because of the increase in transport coefficients with increasing temperature, Reynolds numbers measured in the burnt gas are smaller than those in the fresh mixture. The consequently increased viscosity leads to greater rates of turbulent dissipation in the burnt gas and has been suggested in some cases to produce laminarization of the turbulence. This effect may occur in sufficiently weak turbulence, but in strong turbulence the fluctuations remain, although their dynamics are affected by the changed viscosity.

There are practical configurations in which the temperature increase enhances production of turbulence. For example, a ramjet flame stabilized by a rod in a combustion chamber has a low-density hot zone with recirculation behind the rod, and the density decrease across the flame spreading outward from this recirculation zone produces a velocity increase in the wake that can lead to turbulence generation through instability of the mean velocity profile. From the curl of (1.2), an equation for the vorticity $\nabla \times v$ can be derived, and in this equation appears the baroclinic torque, $\nabla p \times \nabla(1/\rho)$; the changes in p and ρ across the flame can then lead to vorticity generation or reduction, depending on the configuration. In general, pressure and gravity fields act differently on the high-density fresh mixture than on the low-density reaction products. Thus there is a wide variety of fluid-mechanical mechanisms by which premixed flames can influence turbulence.

Turbulent Diffusion Flames

Diffusion flames affect turbulent flows by many of the same mechanisms operative for premixed flames. For example, the diffusion flame in a turbulent mixing layer between fuel and oxidizer streams steepens the maximum gradient of the mean velocity profile somewhat through effects of the baroclinic torque, operating mainly in the wings of the large vortices (where pressure and density gradients are not colinear) and generating vorticity of opposite signs on opposite sides of the high-temperature (low-density) reaction region. The decreased overall density of the mixing layer with combustion increases the dimensions of the large vortices and reduces the rate of entrainment of fluid into the mixing layer. The effects of the combustion can be seen most dramatically in turbulent-jet diffusion flames, in which a high-speed, turbulent, fuel jet issues into a slow or quiescent oxidizing atmosphere. Visualization techniques, employing seeding and laser illumination, have recently shown many details of these jet flows, revealing locations of vortices and flames. Near the jet exit, the turbulence scales are much smaller and the intensities much higher if the flame is extinguished than they are in the presence of the flame. These fluctuations in diffusion flames (and in premixed flames as well) are responsible for turbulent production of noise, which has quite different properties depending on whether or not the flame is present. Noise has been studied for practical reasons (reducing combustion roar) and for basic reasons (as a possible signature for inferring properties of the turbulent combustion). Influences of noise fields on the turbulence and the combustion tend to be weak, except possibly in the turbulent diffusion flames in supersonic combustion (under development for hypersonic propulsion of the "aerospace plane"), where turbulent pressure fluctuations of the same order of magnitude as the mean pressure may occur.

Since diffusion flames do not have burning velocities (Chapter 3), studies of turbulent diffusion flames address other aspects of the influence of the turbulence on the combustion. Peters (1984) and Bilger (1989) give reviews of the subject. There is interest, for example, in knowing profiles of the average temperature and chemical composition, as well as profiles of averages of fluctuations of these quantities for both premixed and diffusion flames. Also for both types of flames, average rates of production of pollutants, such as oxides of nitrogen, are of interest, as are average rates of heat release per unit volume. Conditions for stabilization of turbulent flames against blowoff in high-speed flows need to be known, as do average flame sizes, such as the length of a turbulent-jet diffusion flame. Interactions of vortices with diffusion flames are of interest (Ashurst and Williams, 1991). The intensity of light emitted by turbulent flames is another quantity of practical concern. The best approaches to obtaining all of these items of information generally differ for reaction-sheet and distributed-reaction regimes.

At very small D_ℓ in the distributed-reaction regime, diffusion-flame combustion becomes like premixed-flame combustion because there is thorough turbulent mixing before the chemistry begins. Therefore approaches suitable for describing premixed flames in the distributed-reaction regime also are suitable for diffusion flames as well; approaches of this kind will be identified in the following section. Just as premixed flames with holes may be expected to occur between $D_k = 1$ and $D_\ell = 1$ in Fig. 5.3, so may diffusion flame sheets be expected to have holes in this regime. Little is known about the characteristics of this intermediate regime for diffusion flames, although some studies, such as investigations of the dynamics of holes in diffusion flame sheets, are beginning to shed some light on the subject. There are indications that an additional parameter becomes important, one that in some sense measures the ratio of the root-mean-square mixture-fraction fluctuation to Z_c of Fig. 3.2 (Bilger, 1989). The limit of large D_k for turbulent diffusion flames is one that can be handled even more easily than the corresponding limit for turbulent premixed flames. Good methods are available for addressing many of the previously identified questions for turbulent diffusion flames in the reaction-sheet regime. It is of interest, therefore, to discuss this regime in greater detail before moving to more difficult questions.

In Chapter 3 it was seen that in the limit of infinite D_k, chemical equilibrium is maintained everywhere, and the mixing process can be described in terms of the mixture fraction. Subject to the previously stated approximations, it was found that all state variables become unique functions of the single mixture-fraction variable Z. Experimental methods for space-time-resolved measurement of compositions and temperatures in turbulent diffusion flames by laser-Raman techniques have progressed so remarkably in recent years that it has become possible to test the predicted relationship to Z experimentally. Examples of experimental results by Magre and Dibble (1988) are shown in Fig. 5.9.

In Fig. 5.9 the solid lines are the theoretical equilibrium relationships between the temperature and Z, and the points are results of individual measurements at various times and positions. Experimental

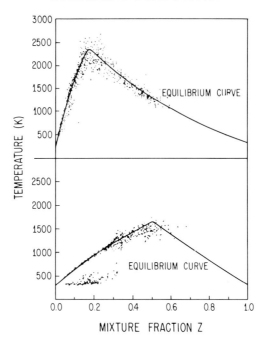

Fig. 5.9. Measured temperatures as functions of the mixture fraction in turbulent-jet hydrogen-air diffusion flames with argon (top) and nitrogen (bottom) dilution of the fuel to different extents.

results for concentrations of individual species as functions of Z are obtained in the same measurements, but to avoid clutter are not presented because they lead to the same conclusions. It must be remembered that the measurement of Z entails measurement of concentrations of all major chemical species and therefore represents a difficult accomplishment, made to appear deceptively simple in Fig. 5.9. These difficulties, combined with problems in achieving sufficiently fine spatial resolutions, lead to some uncertainties in experimental results, although there are appreciably fewer uncertainties for hydrogen flames than for hydrocarbon flames. Continuing advancement of the experimental methods therefore is warranted for improving confidence in results.

The top curve in Fig. 5.9 corresponds to a condition of low jet velocity, leading to low strain rates (small $|\nabla Z|$) and large D_k, so that the equilibrium relationships should be best; although full equilibrium should not be expected, a sufficient number of partial equilibria should give the same result. It is seen that the experimental measurements here cluster well about the equilibrium curve over the entire range of Z for which data can be acquired. This supports the equilibrium relationship. The data, in fact, may be accurate enough here to attribute significance to the spread about the curve, for example to the points showing temperatures above the adiabatic flame temperature, which would not be present if Lewis numbers were unity. Further study may reveal finer details of the flame structure that were previously unanticipated.

The lower curve in Fig. 5.9 has greater nitrogen dilution and a higher jet velocity, both of which bring the system closer to the conditions predicted in Chapter 3 for extinction of the laminar diffusion flames. It is interesting to see that at the lower Z in this figure the data cluster around two bands, one for the equilibrium diffusion flame and the other for the extinguished flame (nonreactive mixing). This is in accord with the theoretical prediction (Chapter 3) of small departures from equilibrium prior to abrupt extinction. Thus the theories of Chapter 3 appear to apply well in this turbulent diffusion flame, at least at sufficiently fuel-lean conditions. For richer conditions there seem to be some intermediate points, but there is some uncertainty in this data; when conditions become rich enough, no data are obtained. Corresponding results for hydrocarbon flames show no evidence for two distinct bands on the fuel-rich side of stoichiometry. This may imply that the ideas of Chapter 3 do not apply to these turbulent diffusion flames, or it may reflect continuing experimental difficulties; more research is needed to clarify the situation.

In any event, the upper curve in Fig. 5.9 supports the hypothesis that there are some turbulent diffusion flames for which the equilibrium limit of large D_k applies and results in an almost unique dependence of all state variables on Z. Although we have called this limit the reaction-sheet regime, in principle the chemistry occurs everywhere to maintain equilibrium and need not be localized in sheets. In practice, however, the reaction-sheet view of this equilibrium limit often is quite good, as may be seen by the sharpness of the peaks of the curves in Fig. 5.9. Some curvature is seen, notably on the fuel-rich side, differing from the straight lines in the reaction-sheet picture in Chapter 3 (Figs. 3.2 and 3.5), but this curvature is small compared with that near the stoichiometric value Z_c. Of course the reaction-sheet picture cannot correlate heavy hydrocarbon and soot profiles in hydrocarbon diffusion flames, but it can apply reasonably well to the major heat-release chemistry. Implications of the equilibrium and reaction-sheet structures on descriptions of turbulent diffusion flames therefore deserve careful attention.

When all state variables in the turbulent flame depend only on Z, it is of great utility to introduce the probability-density function $P(Z)$, such that $P(Z)dZ$ is the probability that the mixture fraction lies in the range dZ about the value Z. The function $P(Z)$ in general varies with position in the turbulent flame, and a method for calculating its evolution therefore is needed. There are various possible approaches to this task, as will be discussed more fully in the following section. Here the essential point is that it is advisable to develop these approaches for the particular quantity $P(Z)$, not for statistical properties associated with other parameters. The fact that the conservation equation for Z does not have a chemical source term greatly reduces uncertainties in approaches for obtaining $P(Z)$. Furthermore, when $P(Z)$ is known, probability-density functions for other variables, for example $P(T)$, are readily constructed from the equilibrium curve. Here let us assume that $P(Z)$ is available and ask what we can calculate on the basis of its knowledge.

Given $P(Z)$, it is straightforward to calculate its average $\overline{Z}$ and averages of its fluctuations, that is, the first two moments of $P(Z)$; for example, its mean and variance are

$$\overline{Z} = \int_0^\infty Z P(Z) dZ \,, \tag{5.26}$$

$$\overline{(Z - \overline{Z})^2} = \int_0^\infty (Z - \overline{Z})^2 P(Z) dZ \,. \tag{5.27}$$

Averages of functions of Z such as $T(Z)$ and averages of their fluctuations are calculable through similar integrals. With these results, the average rate of production of NO can be obtained if the Zel'dovich mechanism ($O + N_2 \rightarrow NO + N$ with the specific reaction-rate constant $k(T)$, followed by the fast step $N + O_2 \rightarrow NO + O$) is applicable. In terms of the mole fractions X_{N_2} and X_O, the molar rate of production of NO is then

$$w_{NO} = 2k(p/R_o T)^2 X_{N_2} X_O \,, \tag{5.28}$$

in which from $T(Z)$ and $X_{N_2}(Z)$, all quantities except X_O are related directly to Z. If O_2 maintains dissociative equilibrium with an equilibrium constant for concentrations $K(T)$, then

$$X_O = \sqrt{X_{O_2}(R_o T/p)K} \,, \tag{5.29}$$

which is thereby related to Z through $X_{O_2}(Z)$ and $T(Z)$; the often-encountered nonequilibrium dissociation leading to super-equilibrium O-atom concentrations requires more detailed consideration of the laminar diffusion-flame structure to find $X_O(Z)$. After the laminar-flame descriptions provide the rate function $w_{NO}(Z)$, its average value locally is simply calculable from

$$\overline{w}_{NO} = \int_0^1 w_{NO} P(Z) dZ . \qquad (5.30)$$

The strong temperature dependences of k and K enable the integral to be approximated accurately by an asymptotic expansion about $Z = Z_c$ and cause $\overline{w}_{NO}$ to be proportional to $P(Z_c)$, the probability density of Z evaluated at $Z = Z_c$.

Calculation of the local average rate of heat release in the reaction-sheet limit requires knowledge of the probability-density function for ∇Z conditioned on $Z = Z_c$. By analogy with (3.2) of Chapter 3 it may be seen that this average rate per unit volume is

$$\overline{w} = -\overline{(\rho D_{12} |\nabla Z|^2 c_p \partial^2 T / \partial Z^2)} , \qquad (5.31)$$

in which arises the instantaneous rate of scalar dissipation,

$$\chi = 2 D_{12} |\nabla Z|^2 . \qquad (5.32)$$

The joint probability-density function $P(\chi, Z)$ thus appears in (5.31), but since in the reaction-sheet approximation $\partial^2 T / \partial Z^2$ vanishes everywhere except for its delta function at $Z = Z_c$, (5.31) can be shown to give for the local average rate of heat release per unit volume

$$\overline{w} = \tfrac{1}{2} \rho_c \overline{\chi}_c P(Z_c) q_F / (1 - Z_c) , \qquad (5.33)$$

where q_F is the heat released per unit mass of fuel consumed. In (5.33) $\overline{\chi}_c$ is the conditioned average rate of scalar dissipation,

$$\overline{\chi}_c = \int_0^\infty \chi P(\chi | Z = Z_c) d\chi . \qquad (5.34)$$

The unconditioned average rate of scalar dissipation $\overline{\chi}$ arises in the conservation equation for $\overline{(Z - \overline{Z})^2}$, derived from (3.11). In the reaction-sheet regime, the turbulent dissipation of the scalar and the process of chemical heat release both occur through molecular diffusion and are both related to statistics of ∇Z. Use of (5.33) again requires knowledge of $P(Z_c)$, but now also the average conditioned dissipation field $\overline{\chi}_c$.

Another aspect of turbulent diffusion flames related to $\overline{\chi}_c$ concerns liftoff and blowoff of turbulent-jet diffusion flames. These phenomena may be described with the aid of Fig. 5.10, which is based on an early

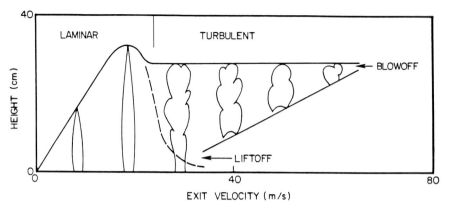

Fig. 5.10. Dependence of the height of a turbulent-jet diffusion flame on jet velocity.

diagram of Hottel and Hawthorne (1949) and illustrates the evolution of the height of a jet diffusion flame for a fixed tube-exit diameter d, as the exit velocity U is increased. In the laminar range, a linear increase of height with U is observed, in accord with Burke-Schumann theory, which predicts the height to be proportional to $d^2 U/D_{12}$. After transition to turbulence the height becomes independent of U but still agrees with the Burke-Schumann proportionality if the laminar diffusivity D_{12} is replaced by a turbulent diffusivity that is proportional to the product dU, as expected by dimensional reasoning. There are intricacies in precise definitions of flame heights both for laminar diffusion flames (is it the height where $Z = Z_c$ on the axis, or the height of the yellow region, or the location of the blue zone on the axis, and how are these related?) and for turbulent diffusion flames (is it the height where $\overline{Z} = Z_c$ on the axis, or should a probability-density function for height be introduced, so that a zero-crossing problem is addressed?) that lead to height differences of roughly 10% by different definitions. Although these intricacies are interesting and in need of further resolution, the general flame-height behavior shown in Fig. 5.10 will remain correct.

The transition to turbulence begins near the top of the flame in Fig. 5.10, and as the exit velocity increases it rapidly moves to the jet exit; at high enough U the turbulence is present everywhere in the flame because the flow within the fuel tube is turbulent. At a critical exit velocity liftoff occurs, that is, the combustion ceases at the jet exit and begins only at an appreciable distance above the exit. Depending on the fuel mixture, liftoff may occur for the laminar flame or may be seen only after the flame has been fully turbulent for a range of U. In either event, as U is increased further after liftoff, the liftoff height h_ℓ, that is, the axial distance from the jet exit to the point at which the combustion begins, increases roughly linearly with U. After h_ℓ has increased to such an extent that it differs from the flame height by an amount comparable with the flame diameter, a further increase in U causes blowoff, that is, the flame is removed and transported downstream, and no further combustion occurs. Turbulent liftoff-height and blowoff behavior can be understood qualitatively from the reaction-sheet viewpoint with account taken of reaction-sheet extinction.

It was seen in Chapter 3 that extinction occurs at a particular value of $|\nabla Z|_c$. According to (5.32), this corresponds to a particular value of χ_c. As $\overline{\chi}_c$ is increased, it may be anticipated that $\overline{w}$ of (5.33) will increase to a maximum value, then extinctions will begin to occur at high-strain positions on the reaction sheet, and so $\overline{w}$ will decrease precipitously. The rate of scalar dissipation χ has the dimension of U/d, and in fact, at the turbulent jet exit, for example, $\overline{\chi}$ is proportional to U/d. Because of the spreading of the turbulent jet, $\overline{\chi}$ decreases with increasing axial distance h from the exit. Measurement and theory for nonreacting turbulent jets show that at sufficiently high Reynolds numbers they develop a self-similar structure in which the integral scale ℓ is proportional to h, and the maximum average velocity (the centerline value) and the maximum fluctuation velocity are proportional to Ud/h, so the turbulence Reynolds number R_ℓ remains constant, ℓ_k/ℓ therefore remains constant, $\overline{Z}$ is proportional to d/h, and $\overline{\chi}$ is proportional to $(U/d)(d/h)^4$. Since Z_c is constant, the self-similar behavior of $\overline{\chi}$ is not exactly reflected in $\overline{\chi}_c$, which is approximately proportional to $(U/d)(d/h)^n$, with $n = 2$ (or possibly n closer to 1). Thus $\overline{\chi}_c$ maintains the constant value corresponding to extinction at a value of h that increases with U, perhaps linearly or in proportion to $\sqrt{U}$. The reaction sheets are mainly extinguished at smaller values of h but can remain largely intact at larger values of h. Therefore, the liftoff height can be approximated as

$$h_\ell = d(U/\chi_e d)^{1/n} , \qquad (5.35)$$

where χ_e is determined from the laminar-flame extinction value of χ_c. With suitable estimates of parameters, (5.35) is in reasonable agreement with experiment (Peters and Williams, 1983).

This reasoning has been based on the reaction-sheet view of turbulent diffusion flames. Lifted diffusion flames have a nonreacting mixing region between the jet exit and the liftoff height. Molecular mixing may occur in this region to such an extent that at h_ℓ the flame in fact is premixed. Reaction-sheet extinctions may be relevant only to the critical velocity U for liftoff and not to h_ℓ. The earliest approaches to calculation of h_ℓ postulated complete mixing and equated the average maximum jet velocity at h_ℓ to the turbulent burning velocity v_T of the stoichiometric mixture. The leading turbulent premixed flame is then followed above by the turbulent diffusion flame in this view. Alternatively, reaction-sheet behavior may occur everywhere in the lifted flame, but the lower end may be composed of edges of diffusion-flame reaction sheets, propagating into a locally nonhomogeneous mixture-fraction field, possibly as a triple flame, with premixed-flame reaction sheets on each side, spreading into ever weaker mixture ratios. In another scenario, partial mixing may occur and be followed by relatively rapid ignition events, motivating detailed study of partially premixed laminar diffusion flames. Further research is needed to better identify conditions under which these various possibilities occur.

Approaches to the Theory of Turbulent Combustion

Methods for attempting to calculate quantities associated with turbulent reacting flows may be divided into the seven categories listed in Table 5.2. It is of interest to discuss the various approaches in the order in which they are listed in the table.

Zero-dimensional and quasi-dimensional methods are approaches in which the spatial characteristics of the turbulence fields are not addressed. The overall effects of the turbulence on processes of interest are approximated, for example through total burning times, or through curves of overall burnt fractions as functions of time, or through curves of mean chamber pressures as functions of time. The effects are parameterized by variables under control of the experimenter, such as mean inlet velocities, spark timing, etc. These methods have been used extensively in the past, for example in design of spark-ignition engines. Although they do not make use of the type of information that has been presented here, they still are applied in industry. With sufficiently

Table 5.2. Approaches to the analysis of turbulent combustion

1. Zero-dimensional and quasidimensional methods
2. Moment methods disregarding probability-density functions
3. Approximation of probability-density functions using moments
4. Calculation of the evolution of probability-density functions
5. Perturbations of known structures possibly extended by renormalization or modeling of the resultant description
6. Field methods not derived from specific fluid-level conservation equations
7. Direct numerical integrations of fluid-level conservation equations

extensive experimental data, these empirical approaches are useful for predicting system behavior through interpolation.

Moment methods that disregard probability-density functions are approaches that have arisen over many years, originally for describing turbulent flows without chemical reactions. They also are called averaging methods because they work with various averages of the conservation equations. They are called moment methods because averages of different powers of flow variables are different moments of the probability-density functions of these variables. Moment approaches formally always lead to systems of partial differential equations that are not closed, in the sense that there are more unknowns than equations. This particular aspect of the closure problem of turbulence arises from the nonlinearity of the conservation equations. Closure problems arise in various forms in almost all of the field-variable approaches to the theory of turbulent combustion. With moment methods, the closure problem is handled by approximating higher moments in terms of lower moments. The assumptions employed in these approximations are based largely on measurements made in turbulent flows (Libby and Williams, 1980).

The level of closure selected in moment methods is a trade-off between the complexity of the calculation and the fidelity of reproduction of experimental results. At a first-order closure level, only the equations for the averages of the flow variables are retained; thus the averages of (1.1) through (1.4) are employed, with all higher moments therein approximated using ideas of effects of turbulence. It has been found that first-order closures fail to retain some memory effects of turbulence. A method that has enjoyed considerable success in recent years and continues to be used extensively is the so-called k-ϵ modeling, in which the differential equations for the averages are supplemented by differential equations for the evolution of the turbulent kinetic energy k and for the average rate of dissipation of turbulent kinetic energy, called ϵ. Even the k-ϵ approach has been found to miss some memory effects, even for nonreacting turbulence. Full second-order closure methods are coming into use; the conservation equations for all of the second moments are retained, and the third moments are approximated. Especially in multicomponent flows, full second-order closures require the solution of relatively large numbers of partial differential equations (as well as the introduction of relatively large numbers of modeling approximations for third and higher moments), but current computer capabilities enable these large numbers of equations to be integrated.

For reacting flows moment methods encounter difficulties that can be illustrated by considering the average of an Arrhenius factor. If k is

Ae^{-E/R_oT}, then

$$\overline{k} = Ae^{-E/R_o\overline{T}} \left[1 + \left(\frac{E}{2R_o\overline{T}} - 1 \right) \frac{E}{R_o\overline{T}^3} \overline{(T - \overline{T})^2} + \cdots \right] . \qquad (5.36)$$

In (5.36) it is seen that the large factor $(E/R_o\overline{T})^2$ multiplies the second moment, and higher moments have higher powers of this large factor. The expansion is convergent, but unless fluctuations are small, successive terms have increasing magnitudes, typically for the first 15 or 20 terms in combustion applications. Therefore (5.36) often is not employed, but instead modeling approximation for the chemical source terms are introduced directly. A successful approximation of this type for ramjet combustion chambers, for example, has been the "eddy-breakup model" (see, for example, Williams, 1985A), in which the chemical sources are approximated in terms of properties of the turbulent dissipation and are not related to the rates of elementary steps (or even to an overall Arrhenius rate).

While this approach has been useful for some applications, it has been found that the resulting approximations interact with approximations for turbulent transport. Turbulent transport terms are usually approximated through turbulent diffusivities to achieve closure, but the chemistry can interfere with this in such a way that there is now both theoretical and experimental confirmation of instances of "countergradient diffusion," that is, of the necessity to introduce negative turbulent diffusivities in portions of the flow to describe reality. This difficulty also arises in methods employing approximations of probability-density functions using moments, but there are indications that it might be overcome by proceeding to full second-order closures. Countergradient diffusion has not been encountered for conserved scalars such as Z, and therefore moment methods for Z seem to be as good as they are for nonreacting flows. Moment methods with full chemistry and without special chemical-source approximations seem relatively well-suited for describing turbulent combustion in the distributed-reaction regime for both premixed and nonpremixed systems. It is in the regime of slow chemistry, with small local relative fluctuations, that these moment methods have achieved their greatest successes. Therefore, despite the difficulties cited, they have a definite place in the theory of turbulent combustion.

Methods involving approximations of probability-density functions using moments focus on probability-density functions for Z or for a reaction-progress variable, for example, and approximate the shapes of these probability-density functions by parameterized functional forms selected to exhibit their anticipated shapes. There may be delta functions at the ends, for example at $Z = 0$ and at $Z = 1$, to account for local portions of the flow in which mixing has not yet occurred (called "intermittency"). A continuous distribution between the boundaries also is included. After the functional form has been selected, the moments are readily computed in terms of the parameters by evaluating integrals, and the parameters are thereby related to the moments. Moment methods are then applied, with partial differential equations integrated to find the evolution of the moments. The number of moments needed is equal to the number of parameters selected to approximate the probability-density functions. From the calculated spatial and/or temporal distributions of the moments, the corresponding profiles of the probability-density functions can then be obtained, and these are used to provide profiles of the flow averages of interest, as has previously been described. The Bilger and Bray articles in the volume of Libby and Williams (1980) expound this type of approach.

These methods are of greatest utility in reaction-sheet regimes. They have been applied to turbulent premixed flames with poor success at first but with improved success as second-order closures have circumvented problems of countergradient diffusion. Their main success, however, has been in the equilibrium limit for turbulent diffusion flames. In this limit, by working with Z the chemical kinetics are removed from the problem entirely. The problem becomes one of turbulent mixing of two nonreacting fluids with complicated thermodynamic and transport properties. It is of interest to illustrate the approach for statistically stationary turbulent flows (for example, turbulent-jet diffusion flames) under the assumption that a two-parameter characterization of $P(Z)$ has been introduced. This illustration will also serve to show some general attributes of moment methods in a little more detail.

It is necessary to have equations for the spatial evolution of $\overline{Z}$ and $\overline{Z'^2}$, where $Z' \equiv (Z - \overline{Z})$. With ρ and D_{12} taken constant for simplicity of illustration, the average of the equation (3.11) for Z is

$$\nabla \cdot (\overline{\boldsymbol{v} Z}) = \nabla \cdot (D_{12} \nabla \overline{Z} - \overline{\boldsymbol{v}' Z'}) , \qquad (5.37)$$

where $\boldsymbol{v}' \equiv \boldsymbol{v} - \overline{\boldsymbol{v}}$. A straightforward calculation with the same equation also gives

$$\overline{\boldsymbol{v}} \cdot \nabla \overline{Z'^2} = -2\, \overline{\boldsymbol{v}' Z'} \cdot \nabla \overline{Z} - \nabla \cdot \left(\overline{\boldsymbol{v}' Z'^2} \right) - 2D_{12} \overline{|\nabla Z'|^2} + D_{12} \nabla^2 \overline{Z'^2} . \quad (5.38)$$

The turbulent-diffusion approximations for the Reynolds-transport terms,

$$\overline{\boldsymbol{v}' Z'} = -D_T \nabla \overline{Z}, \quad \overline{\boldsymbol{v}' Z'^2} = -D_T \nabla \overline{Z'^2}, \qquad (5.39)$$

may be introduced, where D_T, the turbulent diffusion coefficient, is approximated as a constant times the turbulent kinematic viscosity, which in turn is taken to be a constant times k^2/ϵ, the ratio of the square of the turbulent kinetic energy to the average rate of dissipation of turbulent kinetic energy. The dissipation term in (5.38) is then approximated as

$$2D_{12}\overline{|\nabla Z'|^2} = D_{12}\overline{Z'^2}/\ell_Z^2 , \qquad (5.40)$$

where ℓ_Z is a dissipation length for the mixture fraction, analogous to the Taylor scale, and is estimated as a constant times $\sqrt{D_{12}k/\epsilon}$. These forms are suitable for use with k-ϵ modeling of the velocity field to obtain a closed description for the turbulent diffusion flame. Higher-order closures can be pursued, if desired, as has been indicated above.

Some fundamental problems remain with this approach, even for turbulent diffusion flames in the limit of chemical equilibrium. In particular, all of the uncertainties associated with moment methods for nonreacting turbulence remain. In addition, influences of the more complicated thermodynamic properties, with large gas density variations, need to be studied. An important line for future fundamental research directed toward describing turbulent diffusion flames through approximations of probability-density functions is to investigate influences of thermodynamic and molecular transport complications on moment-method modeling. With variable density, mass-weighted (termed Favre) and volume-weighted averages differ, and although the former is now generally preferred because it gives fewer inertial and convective terms in moment equations, further investigation of effects of variable properties on its use is needed. It is worth outlining a new formulation (Liñán, 1991) for diffusion-flame combustion in the reaction-sheet approximation that can help in addressing some of these questions.

The derivation of (3.11) involved the assumption of equal diffusion coefficients for fuel and oxidizer, which often is poor for heavy (large hydrocarbon) or light (hydrogen) fuel molecules. To generalize the reaction-sheet formulation for differing diffusivities, let ρD_{12} denote λ/c_p, the product of the density with the thermal diffusivity, and for the Lewis numbers of Table 1.6 introduce L_F as that of fuel on the fuel side of the reaction sheet and L_O as that for oxidizer on the oxidizer side. Because of the importance of diffusion, it becomes convenient, in generalizing the definition of the mixture fraction for this situation (in which more than one mixture fraction exists as a consequence of the differing diffusivities), to employ fuel and oxidizer mass fractions, each weighted inversely by its Lewis number, in defining the variable Z. If Z_c is the more conventional stoichiometric mixture fraction, obtained from the fuel and oxidizer without the Lewis-number weighting, then the stoichiometric value of the Z with the Lewis-number weighting is

$$Z_s = Z_c / [Z_c + (L_O/L_F)(1 - Z_c)] \ . \tag{5.41}$$

We let L denote L_O for $Z < Z_s$ and L_F for $Z > Z_s$ and thereby derive from (1.1), (1.4), and (1.8)

$$\rho L (\partial Z/\partial t + \boldsymbol{v} \cdot \boldsymbol{\nabla} Z) = \boldsymbol{\nabla} \cdot (\rho D_{12} \boldsymbol{\nabla} Z) \ , \tag{5.42}$$

in which L_F and L_O have been assumed constant. Considering c_p constant and introducing most of the additional approximations stated before (3.2), we can further derive from (1.3) an equation for a nondimensional excess enthalpy H, namely

$$\rho (\partial H/\partial t + \boldsymbol{v} \cdot \boldsymbol{\nabla} H) + \rho N (\partial Z/\partial t + \boldsymbol{v} \cdot \boldsymbol{\nabla} Z) = \boldsymbol{\nabla} \cdot (\rho D_{12} \boldsymbol{\nabla} H) \ , \tag{5.43}$$

where $N = (1 - L_O)/Z_s$ for $Z < Z_s$ and $N = (L_F - 1)/(1 - Z_s)$ for $Z > Z_s$. Oxidizer and fuel mass fractions and temperature are related to Z and H according to

$$\left. \begin{aligned} Y_O &= Y_{Oo}(1 - Z/Z_s) \\[4pt] L_F c_p (T - T_o)/(q_F Y_{Fo} Z_s) &= H + Z/Z_s \end{aligned} \right\} \quad Z < Z_s$$

$$\left. \begin{aligned} Y_F &= Y_{Fo}(Z - Z_s)/(1 - Z_s), \\[4pt] L_F c_p (T - T_o)/(q_F Y_{Fo} Z_s) &= H + (1 - Z)/(1 - Z_s), \end{aligned} \right\} \quad Z > Z_s \tag{5.44}$$

where Y_{Oo} is the oxidizer mass fraction in the oxidizer stream, Y_{Fo} is the fuel mass fraction in the fuel stream, T_o is the temperature of the oxidizer stream, and q_F is as in (5.33).

When the reaction-sheet approximation is good but Lewis numbers differ from unity, equations (5.42) and (5.43) can be used in place of (3.11) as a starting point, since (5.44) serves to relate the variables of interest to this Z and H. For these equations, generalizations of the modeling like that in equations (5.37) through (5.40) can be sought, or direct numerical simulations (as discussed below) can be attempted. It may be noted that, because of the H field, T at $Z = Z_s$ may now differ from the adiabatic flame temperature. These equations can, of course, be used for laminar as well as turbulent diffusion flames. In applications it is often most convenient to have the equations in a so-called "conservation form," in which no variables appear outside derivatives.

This can be achieved by introducing two new explicit functions of Z, namely $J = L(Z - Z_s)$ and $K = N(Z - Z_s)$, to write (5.42) and (5.43), respectively, as

$$\partial(\rho J)/\partial t + \nabla \cdot (\rho v J) = \nabla \cdot (\rho D_{12}\nabla Z) \qquad (5.45)$$

and

$$\partial [\rho(K + H)]/\partial t + \nabla \cdot [\rho v(K + H)] = \nabla \cdot (\rho D_{12}\nabla Z). \qquad (5.46)$$

Further study of these equations could improve our ability to describe turbulent diffusion flames.

The fourth entry in Table 5.2 stems from the fact that turbulent combustion can be viewed as a stochastic process governed by (1.1) through (1.4). For such a process, a probability-density functional can be introduced, which is a functional of all of the velocity and state variables that appear in these equations; a linear functional differential equation for this probability-density functional is readily derived from (1.1) through (1.4). This equation is complete and contains all of the statistical information about the flow. From this equation, simpler equations having less statistical information may be derived by integration over the variables that are not of interest. These simpler equations are not complete, thereby reflecting the closure problem at a fundamental level. However, it is necessary to work with them because methods are not available for solving the functional differential equation. The procedure is to derive partial differential equations for specific probability-density functions by integration, with the probability-density functions chosen to provide the desired information but to depend on a small enough number of variables for their computer solution to be tractable, and to introduce closure approximations in these equations to make the number of dependent variables equal to the number of equations. A formulation in this category is presented in detail by Pope (1985).

To illustrate the result we consider an isothermal, constant-density, constant-diffusivity, reacting mixture of two chemical species with mass fractions Y_1 and Y_2 in an inert carrier gas with the velocity field v specified deterministically. The equation for evolution of the joint probability-density function $P(Y_1, Y_2)$ is found to be

$$\frac{\partial P}{\partial t} + v \cdot \nabla P + \frac{1}{\rho}\left[\frac{\partial(w_1 P)}{\partial Y_1} + \frac{\partial(w_2 P)}{\partial Y_2}\right] = (D_1 + D_2)\nabla^2 P$$

$$-D_1\frac{\partial^2}{\partial Y_1^2}\int_{-\infty}^{\infty} X_{11}P(X_{11}, Y_1, Y_2)dX_{11} - (D_1 + D_2)\frac{\partial^2}{\partial Y_1 \partial Y_2}\int_{-\infty}^{\infty} X_{12}P(X_{12}, Y_1, Y_2)dX_{12}$$

$$-D_2 \frac{\partial^2}{\partial Y_2^2} \int_{-\infty}^{\infty} X_{22} P(X_{22}, Y_1, Y_2) dX_{22} \,, \qquad (5.47)$$

where w is the chemical production rate of the species, D its diffusion coefficient in the carrier, and X_{ij} is the dot product of the gradients $(\nabla Y_i) \cdot (\nabla Y_j)$. There is an analogy here with the Liouville and Boltzmann equations of the kinetic theory of gases. The closure problem lies in the molecular diffusion effects, which involve the joint probability-density function $P(X_{ij}, Y_1, Y_2)$ that includes the spatial gradients. Approximations for these joint probability-density functions have been based on ideas resembling those employed in moment-method closures. The chemical kinetics involve only one-point quantities and are seen in (5.47) to impose no direct difficulties because they are functions only of Y_1 and Y_2, which are independent variables. Nevertheless, the chemical kinetics might insidiously influence appropriate closures for the molecular transport; this could underlie the absence of an influence of reaction-sheet evolution in (5.24).

From (5.47) it is seen that approaches involving integration of equations of evolution for probability-density functions entail solving differential equations in a larger number of dimensions. Although this has restricted their range of application in the past, computer capabilities have developed sufficiently that some problems of practical interest can be addressed; stationarity and strong spatial homogeneity no longer are essential because of advances in efficient integration techniques. The most significant difficulties therefore appear to lie in the closure and in the influences of boundary conditions. Concerning the latter, the statistics of turbulence experimentally are relatively insensitive to small changes in boundary conditions (that is, turbulence possesses a strange attractor), so this same property will have to result from solutions to equations like (5.47). Pursuit of methods of evolution of probability-density functions is a challenging intellectual exercise that deserves more attention than can currently be offered by the few specialists in the subject.

The fifth entry in Table 5.2 is a method that avoids all closure approximations by use of formal perturbation expansions about known structures. This approach has been applied, for example, to premixed turbulent flame propagation in the weak-turbulence limit (Clavin, 1985). Expansions are developed in the ratio of the laminar-flame thickness to the turbulence scales as a small parameter, and often also in the ratio of the turbulence intensity to the laminar burning velocity as another small parameter. With correctly performed expansions, the results are correct, for the range over which the expansion applies. The drawback is that the expansions typically are restricted in their range applicability and exclude many turbulent-combustion flows of practical interest. Sometimes behaviors away from the range of applicability of the expansion can be inferred by studying higher-order terms, but seldom with very good accuracy. More speculative extensions of the results to regions far from the initial small perturbation, as through renormalization, can give potentially more useful predictions that unfortunately also are more controversial. The perturbations do provide accurate limiting results that can be used as tests of performance of other methods. However, even the perturbation methods can be tricky, in that their predictions can depend on uncertain hidden assumptions such as the existence of Fourier transforms.

Field methods not derived from specific fluid-level conservation equations are methods that address the particular problems of interest in a phenomenological way, attempting to calculate the spatial and/or temporal evolutions through *a priori* conceptual models. Examples include "age theories" for chemical reactors and the ESCIMO approach of Spalding (1976), an acronym for engulfment, stretching, coherence, interdiffusion, and moving observer, in which parcels of fluid called "folds" are identified, having both "demographic" and "biographic" attributes, the former being described probabilistically and the latter deterministically through chemical histories within a fold. In age theories a probability-density function for reactant concentration (for example) is introduced, and an evolution equation for it is hypothesized, in which appears a residence time in the reactor. The residence time is ascribed a probability-density function, the age distribution, and based on assumed functional forms of this distribution, histories of reactant concentrations are calculated; from the results, tracer experiments can be proposed for measuring the age distribution (Pratt, 1976). Other methods in this category are procedures for simulation of premixed turbulent-flame propagation by assigning a probabilistic description of forward-jumping of the flame through eddy turnover while retaining a deterministic description of the local propagation of each front, and approaches in which fractal characteristics of reaction sheets are hypothesized and assumed to evolve according to guessed propagation rules. Methods in this general category can turn out to be useful for addressing specific questions of interest that are difficult for other theories to approach, but they also can represent dead-end streets when their ad hoc bases preclude tie-ins with other approaches.

With the increasing computer capabilities, direct numerical integrations of fluid-level conservation equations, also called turbulence simulations, are mushrooming as methods for studying turbulent reacting flows. Many approaches in this category have been developed and applied, from solution of relatively straightforward finite-difference approximations to the differential equations, to spectral methods in which transforms are introduced and the equations are solved in the transform space then the solutions converted back to the original space, to random-vortex methods in which the flow is discretized into vortices that interact with each other and are tracked by the computer algorithm (Majda, 1991). The most efficient simulation approach depends on the particular problem; finesse in execution is needed to make any of these methods work well. In the long run, these simulation methods will not require closure approximations and will demonstrate the long-time large chaotic departures of adjacent fluid elements from each other, with strong sensitivities of these departures to initial and boundary conditions but correspondingly weak sensitivities of the statistics, which are properties of the deterministic equations, as with all strange attractors. In the short run, subgrid-scale closure approximations, such as that of Kerstein (1991B), still are needed for the large turbulence Reynolds numbers of interest. Intensive research may be expected to continue on closures for these large-eddy simulation techniques and to show new aspects of turbulent reacting flows.

The needed computational capacities increase so rapidly with the number of grid points that specialized computers, for example a future "turbulent-combustion computer" with parallel architecture, are attractive for hastening the availability of results for high Reynolds numbers. Even today the numerical outputs possess a tremendous amount of information retrievable more readily than from experiment. Questions arise as to what information to select. The simulations do not help in making this selection; theories based on different approaches point out the relevant items to extract from the simulations to enhance our understanding. The development of numerical-integration methods may be viewed as constructing a laboratory in which "experiments" can be run to address the questions that theories suggest are important.

Outstanding Problems in Turbulent Combustion

In the preceding discussions we have indicated a variety of questions in turbulent combustion that remain to be resolved. Table 5.3 is a list of a few selected problems that are in need of further clarification. The approaches identified in the preceding section can help in addressing these problems. Table 5.3 will be discussed further in the following chapter. The following bibliographical citations are mainly drawn from

Table 5.3. Outstanding problems in turbulent combustion

1. The dynamics of premixed flame surfaces in turbulent fields in the reaction-sheet regime
2. The modifications to turbulence caused by premixed flames in the reaction-sheet regime
3. Structures of turbulent flames in regimes in which reaction sheets exist but are partially disrupted by turbulence
4. Premixed turbulent flame thicknesses and burning velocities
5. Blowoff conditions for turbulent flames
6. Average rates of heat release in turbulent combustion
7. Lengths of turbulent flames
8. Intensities of radiant energy emitted by turbulent flames
9. Liftoff conditions and liftoff heights of turbulent jet diffusion flames

reviews and more recent research publications that provide entries into the extensive literature.

Bibliography

Aldredge, R.C. and Williams, F.A. 1991. *J. Fluid Mech.* **228**, 487.

Ashurst, W.T., Kerstein, A.R., Kerr, R., and Gibson, C.H. 1987. *Phys. Fluids* **30**, 2343.

Ashurst, W.T. and Williams, F.A. 1991. *Twenty-third symposium (international) on combustion.* The Combustion Institute, Pittsburgh, pp. 543–550.

Ballal, D.R. and Lefebvre, A.H. 1975. *Proc. R. Soc. London* **344A**, 217.

Bilger, R.W. 1989. *Twenty-second symposium (international) on combustion.* The Combustion Institute, Pittsburgh, pp. 475–488.

Chomiak, J. 1990. *Combustion: A study in theory, fact and application.* Gordon and Breach, New York, Chapter 3.

Clavin, P. 1985. *Prog. Energy Combust. Sci.* **11**, 1.

Damköhler, G. 1940. *Z. Elektrochem.* **46**, 601.

Goulard, R., Mellor, A.M., and Bilger, R.W. 1976. *Combust. Sci. Technol.* **14**, 195.

Hottel, H.C. and Hawthorne, W.R. 1949. *Third symposium on combustion and flame and explosion phenomena.* Williams and Wilkins, Baltimore, pp. 254–266.

Karlovitz, B., Denniston, Jr., D.W., Knapschaefer, D.H., and Wells, F.H. 1953. *Fourth symposium (international) on combustion.* The Combustion Institute, Pittsburgh, pp. 613–620.

Kerstein, A.R., Ashurst, W.T., and Williams, F.A. 1988. *Phys. Rev. A.* **37**, 2728.

Kerstein, A.R. 1988A. *Combust. Sci. Technol.* **60**, 163.

Kerstein, A.R. 1988B. *Combust. Sci. Technol.* **60**, 441.

Kerstein, A.R. 1991A. *Phys. Rev.* **A44**, 3633.

Kerstein, A.R. 1991B. *J. Fluid Mech.* **231**, 361.

Kerstein, A.R. and Ashurst, R.C. 1992. *Phys. Rev. Lett.* **68**, 934.

Klimov, A.M. 1975. *Dokl. Akad. Nauk SSSR* **221**, 56.
Klimov, A.M. 1983. *Prog. Astronaut. Aeronaut.* **88**, 113.
Kuznetsov, V.R. and Sabelnikov, V.A. 1986. *Turbulence and combustion.* Nauka, Moscow; English translation, 1990, Hemisphere Publishing Corporation, New York.
Libby, P.A. and Williams, F.A., eds. 1980. *Turbulent reacting flows.* Springer-Verlag, Berlin.
Libby, P.A., Williams, F.A., and Sivashinsky, G.I. 1990. *Phys. Fluids* **A2**, 1213.
Libby, P.A. and Williams, F.A., eds. 1993. *Turbulent reacting flows.* Academic Press, London.
Liñán, A. 1991. *El papel de la mecánica de fluidos en los procesos de combustión.* Real Academia de Ciencias Exactas, Fisicas y Naturales, Madrid.
Magre, P. and Dibble, R. 1988. *Combust. Flame* **73**, 195.
Majda, A.J. 1991. *SIAM Review* **33**, 349.
Mantzaras, J., Felton, P.G., and Bracco, F.V. 1989. *Combust. Flame* **77**, 295.
Markstein, G.H. 1964. *Non-steady flame propagation.* MacMillan, New York.
Murayama, M. and Takeno, T. 1989. *Twenty-second symposium (international) on combustion.* The Combustion Institute, Pittsburgh, pp. 551–559.
Peters, N. and Williams, F.A. 1983. *AIAA J.* **21**, 423.
Peters, N. 1984. *Prog. Energy Combust. Sci.* **10**, 319.
Peters, N. 1987. *Twenty-first symposium (international) on combustion.* The Combustion Institute, Pittsburgh, pp. 1231–1250.
Peters, N. and Williams, F.A. 1989. *Twenty-second symposium (international) on combustion.* The Combustion Institute, Pittsburgh, pp. 495–503.
Peters, N. 1992. *J. Fluid Mech.* **224**, 611–629.
Pope, S.B. 1985. *Prog. Energy Combust. Sci.* **11**, 119–192.
Pope, S.B. and Anand, M.S. 1985. *Twentieth symposium (international) on combustion.* The Combustion Institute, Pittsburgh, pp. 403–410.
Pratt, D.T. 1976. *Prog. Energy Combust. Sci.* **1**, 73.
Sivashinsky, G.I. 1983. *A. Rev. Fluid Mech.* **15**, 179.
Sivashinsky, G.I. 1988. *Combust. Sci. Technol.* **62**, 77.
Spalding, D.B. 1976. *Combust. Sci. Technol.* **13**, 3.
Williams, F.A. 1985A. *Combustion theory*, 2nd ed. Addison-Wesley Publishing Company, Menlo Park, California.
Williams, F.A. 1985B. Turbulent combustion. In *The mathematics of combustion*, J.D. Buckmaster, ed., vol. 2 of *Frontiers in Applied Mathematics*, Society for Industrial and Applied Mathematics, Philadelphia, pp. 97–131.
Yakhot, V. 1988. *Combust. Sci. Technol.* **60**, 191; **62**, 127.

6
THE FUTURE

The preceding chapters have offered a survey of the breadth and depth of the field of combustion. With this background it is tempting to try to predict what the future will bring to the subject. Historically, predictions have been notably inaccurate, especially for periods exceeding five years. There are tendencies to be overoptimistic about what can be accomplished in five years and especially overpessimistic about prospects for advances over much longer time periods. Totally unanticipated developments (witness the recent breakthrough in superconductivity) can entirely reverse projections for future progress. With these reservations in mind, some observations will be offered here on the possible future of combustion.

Energy Sources and Fire and Explosion Safety

From a practical aspect, we may guess that overall interest in the field will remain approximately at the same high level that it enjoys today. After the advent of atomic energy, combustion sometimes was predicted to die as a source of heat and power production. The well-known pollution problems of smoke, oxides of sulfur and of nitrogen, carbon monoxide, hydrocarbons, particulates, and diesel odor were thought to be avoidable entirely by clean and abundant atomic energy. It was found that, however, fission has numerous pollution problems of its own, and the energy available in the earth's deposits of fission fuels is no greater than that available in its storehouse of fossil fuels of combustion. Fusion, in principle, through hydrogen isotopes offers by many orders of magnitude the most abundant energy supply available to the earth, but practical utilization of such fuels in significant proportions by hot or cold fusion seems generations away and could well encounter unique safety and pollution problems that would render combustion still attractive. Interest in combustion for heat and power seems likely to remain at its current level, at least in our lifetimes—and those of our grandchildren.

Even if combustion as a source of power were to decline, practical interest in the subject would remain in connection with safety and fire hazards. The continual invention of new construction materials poses an evolving need for investigation of their flammability hazards in urban fires. Most synthetic materials have appreciably higher heats of combustion than wood and therefore contribute greater fuel-energy loading in unwanted fire situations. Wildland fires are unavoidable—in fact, often ecologically essential—and therefore are expected to remain a topic of study. Alternative power sources, such as nuclear energy, are found to exhibit unique hazards of fires and explosions through combustion. Thus, in the area of safety, combustion always will be a concern.

The energy supplies in fossil fuels are mostly in the form of coal, largely in the United States, the USSR, and China. The coal supplies appreciably exceed those of petroleum and natural gas. For many applications, the latter two are more desirable because of their relative ease of transportation by pipeline. Liquid fuels are so advantageous for mobile power (automobiles, airplanes) that there is incentive to maximize their availability. Prospects are good for future development of new sources of liquid fuels. For example, oil shale, a relatively untapped source of liquid fuel (shale oil), is available in abundances comparable with those of petroleum. Techniques for extraction of this resource, especially by in situ methods, may be expected to advance greatly in the future, whenever the economic aspects become favorable (and surely they will, at some time). In addition, methods for coal liquefaction, as well as coal gasification, may be anticipated in the future to begin to contribute significantly to our liquid and gaseous fuel supplies. Associated environmental problems are substantial but not necessarily greater than those associated with comparable levels of utilization of alternative energy sources. Renewable energy resources in vegetation, such as wood, may play more prominent roles in fulfilling fuel needs but are unlikely to be sufficiently developable to replace fossil fuels as the major contributor. With these various sources of energy for combustion, mankind can be expected to continue to enjoy its benefits for many generations.

Aerospace Propulsion

Figure 1.2 is an illustration of what can be accomplished in the technology of propulsion through use of combustion. It is safe to predict that combustion will continue to contribute to future advances in aerospace propulsion. In the order of decreasing efficiency and thrust, photon, ion, electrothermal, and nuclear engines are all attractive for deep-space propulsion. However, for all but the last of these, which raises monumental safety and environmental issues, trips originating from the surface of the earth would need auxiliary motors. Propulsion systems employing combustion are promising not only for flight from earth into orbit but also for manned space flight throughout the solar system and flight in atmospheres of other planets.

Will the future aerospace propulsion systems based on combustion employ solid, liquid, or gaseous fuels? Societal as much as technological considerations will dictate the answer. The space shuttle employs both solid-propellant and liquid-propellant rocket motors, at least partially because each of these industries has a significant political constituency. The space shuttle does not make use of air-breathing propulsion while in the earth's atmosphere, even though paper studies show significant potential improvements in performance associated with the consequent removal of the necessity of carrying an oxidizer from the earth's surface. We predict that in a ten-year time frame the strong technological challenge of developing a workable airbreathing engine system for aerospace propulsion will be solved and that this system will employ supersonic combustion of hydrogen in air. Recent research has in fact significantly increased our understanding of supersonic hydrogen-air combustion. The earth-to-orbit aerospace plane may thus become a reality with gaseous or liquid hydrogen as its fuel. However, we also predict that in a longer-term time frame, upwards of twenty years, the much greater volumetric energy densities and handling conveniences of non-cryogenic liquid fuels will lead to the emergence of these liquids as the primary propellants—especially energetic hydrocarbon-based liquids.

Predictions of Specific Advances in Combustion Science

Having truly entered the realm of speculation, we now return to the tables at the end of each preceding chapter to predict where the future scientific advances will occur. It is convenient to consider first laminar combustion (Chapters 2 and 3), then ignition and detonation (Chapter 4), and finally turbulent combustion (Chapter 5). After these discussions of specifics, we shall proceed in the next section to less speculative and more general observations on the three principal scientific components of the subject—experiment, computation, and theory.

Entries 1 in Tables 2.5 and 3.3 concern laminar flame structures with chemistry that is more complicated than the simple one-step Arrhenius process. Most of the research on model chemistry for hypothetical systems, selected to simulate behaviors of real systems, is now completed, insofar as development of qualitative understanding is concerned, and its further use will be for quantitative predictions of specific models empirically tied closely to real chemistry. The main future tasks are related strongly to real kinetics. Numerical integrations with full chemistry in one dimension have become routine, and we are well on our way to achieving a comparable status in two dimensions. We predict that within the next ten years, two-dimensional, time-dependent numerical computations with full chemistry will become available as standard tools for combustion scientists. With current computational methods, it is more difficult to calculate flame structures with highly reduced chemical-kinetic mechanisms than with full mechanisms. A paramount outstanding problem for the numerics, which we predict will be solved in five years, is the development of robust solvers of the complicated systems of algebraic equations that describe chemical-kinetic steady states, to be meshed with existing flame-structure programs, enabling flame structures to be calculated easily with any degree of reduction of the kinetic scheme. On the side of theory, RRA is rapidly becoming dominant and in the next five years may be expected to be responsible for great advances in the understanding of structures of real flames—not only hydrocarbon-air, hydrogen-oxygen, and hydrogen-halogen flames, but also carbon-monoxide flames and flames of propellants involving nitrogen chemistry, as well as processes of pollutant production in flames. The methods of AEA and DNA will continue to play important roles in these studies, but RRA will take the lead.

The developing two-dimensional and three-dimensional capabilities will enable significant progress on item 2 in Table 2.5 to be made. The theoretical approaches to describing influences of curvature and strain with real kinetics will make strong use of perturbation methods. The flame stability studies identified as items 3 and 4 in Table 2.5 and as item 4 in Table 3.3 also must emphasize real kinetics, simplified, for example, through RRA. Our current understanding of instabilities of premixed flames and of diffusion flames is based almost entirely on concepts from one-step, Arrhenius chemistry. The extents to which these concepts extend to real chemistry will be clarified greatly during the next five years. Corresponding clarification may be anticipated concerning flame extinction in gaseous systems with real kinetics, identified in entries 5 of Table 2.5 and 2 of Table 3.3.

The last two premixed-flame topics in Table 2.5 are subjects that do not specifically emphasize real kinetics. They are example problems from flame dynamics that necessitate considerations of nonadiabaticity and of cellular flames. There are other examples deserving further study as well, such as flame balls (Ronney, 1990), in which, for reasons that are partially understood, stationary, spherical, premixed flames may sit in an infinite, quiescent medium. The last topic in Table 2.5 is of interest to lean-burning or dual-chamber spark-ignition engines in which pollutant reduction is sought through enthalpy variation; it raises questions about the possibility of flames propagating dynamically into regions where the mixture lies outside the usual flammability limits. These topics cannot be addressed well by perturbation methods. Bifurcation theory is of limited use because the interest lies far from a bifurcation point. Although experimental and computational approaches can be employed, it will be difficult to derive a theoretical understanding from such investigations. Therefore only slow progress is predicted for items 6 and 7 of Table 2.5.

Entry 3 of Table 3.3 identifies a subject of practical interest for both liquid-fuel combustors and fire extinguishers. Idealized configurations such as counterflows have been exploited very little for these two-phase systems. During the next five years, increasing experimental, computational, and theoretical investigations of such configurations are anticipated, leading to improved understanding of spray combustion and of flame extinguishing by dry-powder fire suppressants.

Currently progress is being made on theoretical analysis of structures of diffusion flames extending to cold boundaries, for example with identification of triple-flame configurations in which curved rich and lean premixed flames meet at the end of a diffusion flame. Item 5 of Table 3.3 thus may be expected to enjoy advancement. However, the problem is a difficult one. Both numerical and analytical methods will be fruitful. Experimentally it would be desirable to establish such nonuniform flames in steady-flow configurations in the laboratory, but it is difficult to do so. Since buoyancy is one of the complicating factors, microgravity experiments in space vehicles could be useful in these studies.

Microgravity also can aid in investigating items 6 and 7 of Table 3.3. Combustion experiments on flame spread over a solid paper (cellulose) fuel sample have in fact been performed in an apparatus in the space shuttle (Bhattacharjee and Altenkirch, 1993). Although no other combustion-science experiments in space have yet been completed, many more can be expected in the next five to twenty years. They could be especially useful for item 7 because by varying the buoyancy its effect can be sorted out from that of gradients of surface tension in driving flows and flames over liquid surfaces. The two-phase character of the problems in 6 and 7 presents difficult challenges to theory, but nevertheless we predict that significant further progress will occur over this same time frame.

What advances may we expect in the next five years in the areas of ignition and explosions? Items 1 and 2 of Table 4.2 are among the few problems in asymptotic analysis in ignition theory for one-step, Arrhenius chemistry that have not yet been solved completely; further advancement may be anticipated here. A related problem (not listed), for which advancement also will occur, is the corresponding theory of ignition in nonuniform mixing regions, which is relevant to Diesel combustion. However, just as for laminar flame structures, the most rapid advancement will occur for non-one-step kinetics (item 3 of Table 4.2), again through numerics and RRA. Clarification of the dependence of ignition on the size and shape of the hot body that causes the ignition (entry 4) is a relatively difficult problem of practical importance that will experience only slow progress. The same may be said for item 5; numerical integrations, perhaps more than analytical approaches, will provide the main vehicle for improvement. The transient development of pressure waves in the explosion, following the initial stages (entries 6 and 7 of Table 4.2), brings in gasdynamics and disparate time scales; careful numerical and analytical approaches will enjoy some success for symmetric configurations (item 6), but little success for configurations without symmetry, item 7, which will continue to have to rely on rough, overall, conservation principles. Explosions in chambers with and without venting will continue to pose safety concerns that motivate basic research.

We predict that scientific investigations of detonation phenomena will proceed but not accelerate in the next ten years. The methods employed will be somewhat less experimental and more computational and analytical. Specific topics that will be studied are listed in Table 4.3. Although not stated in the table, here too more consideration will be given to real kinetics. The first entry identifies analytical approximations to be obtained from simplifications of the chemistry. Associated with numerical and analytical studies of detonation stability (the second and third entries) will be corresponding calculations of nonlinear cellular detonation dynamics. Direct initiation of both strong, stable waves and cellularly unstable waves (items 4 and 5) will receive only modest further study. More attention will be paid to questions related to detonation failure, such as entries 6 and 7. These last two topics are not only of practical importance, they are also challenging to theory and amenable to further improvements in understanding. Our comprehension of conditions for detonation failure will increase appreciably.

Turbulent combustion will remain a subject of strong scientific interest. Experiment, computation, and theory will all play important roles here. Advancement in computational capabilities is especially critical in turbulent combustion. Parallel architecture could be helpful because of the repetitive nature of the calculations. We predict that within twenty years such a turbulent-combustion computer will become available, but that even it will be incapable of reaching an integral-scale Reynolds number R_ℓ of 10^4 (see Fig. 5.3) in three dimensions. Large-eddy simulations with subgrid-scale modeling therefore will become the preferred method of computation for combustor design. Intensive activity and impressive developments in subgrid-scale modeling for turbulent combustion is predicted over the next ten-year period. These are essential if large-eddy simulation is to become useful.

Table 5.3 lists specific problems in turbulent combustion for future study; others are identified in the text in Chapter 5. Entries 1 through 5 of Table 5.3 pertain to premixed turbulent flames, and entries 5 through 9 concern turbulent diffusion flames. Items 1 and 2 are sure-fire bets for advancement in the next five years. Item 3 is much more difficult; yet, we predict considerable success here, over a ten-year period. Item 4 will remain the fundamental scientific testing ground for competing theories of premixed turbulent combustion. Item 5 poses difficult problems, and further understanding of turbulent-flame blowoff from stabilizing rods, tubes, etc., will advance only slowly. Probabilistic methods, coupled with improved understanding of flow structures and advancement of experimental instrumentation, will engender significant progress concerning turbulent-flame heat-release rates and flame geometry (items 6 and 7). We predict significantly improved understanding, over a ten-year time scale, through theoretical and experimental methods, not only for radiant energy emissions from turbulent flames (entry 8), but also for emissions of pollutants from turbulent diffusion flames. Finally, our understanding of liftoff (item 9) and of the relative importance of premixed flames and diffusion flames in determining liftoff heights will advance appreciably over this same time period, again through theoretical and experimental techniques. These last two topics (8 and 9), in our estimation, are too difficult for computation to make a significant contribution in turbulent combustion at high Reynolds numbers.

General Considerations in the Science of Combustion

From a scientific viewpoint, future progress in combustion may be expected to benefit significantly from continuing technological advances. In the early part of this century, increases in understanding of combustion phenomena were achieved mainly by scientists working alone or in pairs, performing experiments in their own small and inexpensive laboratories that they had designed, or mulling over particular simplified conservation equations that appeared to them to describe a poorly understood combustion process. The contribution of the individual scientist, working with meager resources, even today is an important source of advancement, and in the future it will continue to be so; there is too great a tendency nowadays to underestimate the continuing impact of the individual contribution. However, there are new realms of investigation of nearly equal importance that involve expensive equipment and the cooperation of many researchers. For example, there are combustion laboratories (such as that at Sandia, Livermore, California) in which laser diagnostics are piped through laboratory walls like electricity, gas, and water. The design and construction of a laboratory complex of this type involves many people from different disciplines. Many do not have very good ideas of what the current outstanding combustion problems are; they cannot, because maintaining top proficiency in their specialization is a full-time occupation. Yet, the availability of such facilities enables measurements to be made that not long ago were impossible. Further developments of this kind seem certain to improve scientific progress in combustion. For example, a combustion facility may be constructed in a space station that will enable investigations to be pursued without the complication of buoyancy; this would require the cooperation of a large number of scientists and engineers. Although it is impossible to predict just what experimental advances will occur, it seems safe to say that in the future large, cooperative, experimental efforts will be a source of significant progress in the field.

Continuing advances in computer speed and memory are sure to have a strong impact on combustion science. For the most part, the equations are known—the uncompleted task is to solve them. The magnitude of the task is so great that it seems safe to predict that full description with complete chemistry of a turbulent combustion process by computer is more than a generation away. However, at some time in the future computational facilities will rival or exceed experimental facilities as laboratories for investigation of combustion phenomena. Movement in this direction already is evident, despite the clear inadequacies of even supercomputers for describing three-dimensional, time-dependent flows with many chemical reactions. Puzzling over the meaning of computer outputs is becoming as much a part of theoretical activity in the field as is searching for explanations of unexpected experimental results. In fact, there seems to be a current tendency to put too much faith in computational abilities for design of combustion devices, without sufficient regard for their limitations. Experiment will always be a part of the science; something of central importance just might not be in the equations, even if they are solved exactly. Computers dedicated to specific tasks—reacting-flow computers—may

be developed to hasten arrival of the day on which complicated reacting flows can be calculated completely. When this capability is achieved,
computer laboratories will enable many aspects of combustion processes
to be extracted that are impossible to measure experimentally. Just as
with the large experimental facilities, progress in this computational area
will require cooperation of scientists in widely differing disciplines.

Chemical kineticists, fluid mechanicists, and applied mathematicians all have contributed greatly to progress in combustion science, the
former from the beginning of the science and the latter two especially
in recent years. Combustion was "discovered" by applied mathematicians about twenty years ago, and seeing the challenges in the equations, they have made it an important part of applied mathematics. The
fluid-dynamic challenges of the subject have been recognized sporad-
ically in the past, and only in perhaps the last five years has the fluid
dynamics of reacting flows become a significant part of fluid mechanics.
Many current combustion problems need the efforts of experts in these
disciplines. Asymptotic methods, bifurcation analysis, statistical methods, renormalization theory, fractals, the theory of chaos, etc., need to
be applied to achieve further advances in understanding of combustion
phenomena. For example, in turbulent combustion inabilities of computation to resolve the smallest scales require the investigation of small-
scale closures through asymptotic concepts. Combustion in vortices, the
fluid dynamics of reacting turbulent flows, supersonic combustion, and
detonation structure are examples of problems in need of more investigation. Contributions can be made not only by those who have had
interest in the subject in the past but also by physicists, chemists, engineers, and mathematicians who thus far have encountered combustion
only in their nonprofessional life. There are so many intricacies in need
of better understanding that combustion is likely to remain an active
science into the distant future.

Bibliography

Bhattacharjee, S. and Altenkirch, R.A. 1993. "A Comparison of Theoretical
    and Experimental Results in Flame Spread Over Thin Condensed Fuels
    in a Quiescent Microgravity Environment." *Twenty-fourth Symposium
    (International) on Combustion.* The Combustion Institute, Pittsburgh, to
    appear.
Ronney, P.D. 1990. *Combust. Flame* **82**, 1–14.
The research journals *Combustion and Flame* and *Combustion Science and Tech-
    nology*, the review journal *Progress in Energy and Combustion Science*,
    and the proceedings of the biennial international combustion symposia,
    published by The Combustion Institute, are primary sources of informa-
    tion on the continuing advancement of combustion research.

INDEX

Acetylene, 66, 68
Activation energy, 8, 10, 14, 15, 23, 28
Adiabatic flame temperature, 17
AEA, 14, 15, 24–32, 69–73, 122, 123, 126
Aerospace plane, 14, 132, 155
Afterburner, 112, 113
Ammonium nitrate, 90
Arrhenius law, 22, 23, 28, 85, 86
Atomization, 58–60
Autoignition, 90–93

Baroclinic torque, 131
Bifurcation, 104, 161
Blowoff, 21, 133, 137–39
Boltzmann number, 10, 11
Branched-chain explosions, 83–85, 95, 96
Branched-chain reactions, 49
Bunsen burner, 16, 21, 34
Buoyancy effects, 33–35, 61, 62, 69, 78, 80, 112, 113, 157
Burke-Schumann diffusion flames, 63–65, 69, 138
Burning rate, 57, 77–80
Burning velocity, 21–23, 26–31, 47, 54, 57, 92

Carbon monoxide oxidation, 50–53
Cellular detonations, 104
Cellular flames, 37, 38, 42
Chain branching, 49, 52, 53, 83–85, 96
Chain carriers, 48, 49, 65, 84

Chain initiation, 48, 84
Chain propagation, 48, 84, 96
Chain reactions, 48, 49, 83–85, 96
Chain termination, 48, 53, 84, 96
Chapman-Jouguet combustion waves, 18–20, 101
Chemical equilibrium, 7, 133
Chemical kinetics, 7–11, 49–51
Chemiluminescence, 14, 68
Cloud burning, 58
Combustion instabilities, 32–38, 103, 104
Condensation shocks, 101
Conservation equations, 7–12. *See also specific type*
Convective-diffusive zone, 24–26, 29, 30, 64–68
Convective-reactive zone, 46
Cool flame, 33, 95, 96
Counterflow flames, 21, 63, 70
Countergradient diffusion, 142

Damköhler number, 10, 15, 43, 63, 68–70, 115–17, 123
Deflagrations, 16–19, 21–55, 99
Detonability limits, 106–8
Detonations, 16–19, 98–109
Diesel engine, 5, 58, 118
Diffusion, 10–13
Diffusion coefficients, 8–13
Diffusion equations, 12, 13
Diffusion flames, 15, 16, 57–73, 132–40, 143–46
Diffusion velocity, 9, 10, 12, 13
Diffusive-thermal instability, 35–38

Diffusivity, 24, 26, 79, 80
Dispersion relations, 35–37
Disruptive burning, 74, 75
Dissociation 45, 48
Distributed-reaction regime, 115–19, 133
DNA, 15, 43
Droplet burning, 74–78
D-square law, 77, 78

Eddy, 114, 125
Eddy turnover time, 115
Eddy-breakup model, 142
Emissivity, 12
Energy conservation, 9–12
Enthalpy, 9
Equations of state, 7, 12
Equilibrium, 7, 10, 44. *See also* Chemical equilibrium
Equilibrium constants, 10
Equivalence ratio, 22
Evaporation constant, 77
Excess enthalpy, 145
Explosion limits, 95–97
Explosion peninsula, 95
Explosion temperature, 95
Explosion time, 85
Explosions, 83–96
Explosives, 6, 15, 91, 106
Extinction, 57–59, 69–73, 88, 89, 121–126

Failure diameters, 107
Favre averaging, 144
Fick's law, 13
Fire suppressants, 69
Flame angle, 21, 119
Flame arrestor, 98
Flame balls, 156, 157
Flame curvature, 121, 122, 126
Flame height, 16, 69, 138, 139
Flame holder, 21, 31, 41
Flame instability, 29–35
Flame shape, 16, 33, 34, 62, 69, 138
Flame sheets, 63, 64, 121, 122, 126
Flame speed. *See* Burning velocity
Flame spread, 78–81
Flame stabilization, 21, 131–33, 138–40

Flame stretch, 121–26
Flame temperature, 17, 22–24, 66, 73
Flame thickness, 23–27, 39, 62, 99, 116, 126
Flame-generated turbulence, 105, 131
Flamelets, 64, 121–23, 133–35
Flames, 6, 14–17. *See also specific type*
Flammability limits, 92, 95–98, 106
Flashback, 21
Flat-flame burner, 21, 34
Fractals, 124, 125
Froude number, 10, 11, 113
Fuel-consumption zone, 52, 53, 67
Fullerenes, 6
Furnaces, 4–6, 57, 58, 118

G equation, 126
Galloping detonations, 103, 104
Gas turbine, 5, 112, 113
Gasification, 79
Gibson scale, 125
Gravity effects, 33, 80. *See also* Buoyancy effects

Heat, 10–12
Heat capacity. *See* Specific heat
Heat conduction, 11
Heat loss, 21, 86–90, 97, 98
Heat of combustion, 8–11
Heat of dissociation, 45, 48
Heat of formation, 9
Heat of reaction, 8–11
Heat of vaporization, 74–77
Heat-flux vector, 9–12
Heat-transfer coefficient, 86, 87
Hugoniot curves, 17–20, 100
Hugoniot equation, 17, 18
Hydrocarbon combustion, 50–52, 66
Hydrocarbon-air flames, 49–55, 66–69
Hydrodynamic instability, 33–35, 130, 131
Hydrogen-bromine reaction, 47–49
Hydrogen-chlorine reaction, 65
Hydrogen-halogen flames, 47–49
Hydrogen-oxygen flames, 49–51

Hydrogen-oxygen reaction, 49, 95, 96

Ideal gas law, 12
Ignition, 59, 61 , 71, 91–94
Ignition delay, 94
Ignition energies, minimum, 91–93, 98
Ignition temperature, 79, 92–94
Ignition theory, 93, 94
Induction time, 102
Inertial subrange, 114
Integral scale, 113–16, 124, 125, 139
Intermittency, 143

Jet-stirred reactor, 118
Jump conditions, 17

Karlovitz number, 122
K-ϵ modeling, 141, 144
Kinematic viscosity, 113, 114, 144
Kinetic theory of gases, 7, 12, 13
Kolmogorov scale, 114, 117
Kolmogorov time, 116

Laminar flames, 21–55, 61–73
Lasers, 6, 40, 61, 129, 132–34, 160
Lewis number, 14, 24–27, 35–38, 121, 122, 144–46
Liftoff, 137–39
Liquid propellants, 58

Mach number, 10, 11, 13, 14, 101–4, 107, 108
Markstein number, 126
Mass conservation, 9, 10
Mass fraction, 9
Mass-flux fraction, 40
Mass-transfer theory, 76–78
Matched asymptotic expansions, 29, 65, 70
Matching, 29
Metal combustion, 58
Metastable state, 26
Methane, 15, 49–55, 66–68, 73
Minimum ignition energy, 91–93, 98
Mixture fraction, 62–64, 75, 76, 133–36, 142–46

Mole fraction, 9
Moment methods, 140–43
Moments, 136
Momentum conservation, 9–12
Multiphase flow, 58–61

Natural gas, 15, 154
Navier-Stokes equations, 7–10
Nitrogen oxides, 59, 136, 137, 153
Nucleation, 74, 75

Oil shale, 154
One-dimensional flow, 17
Order of reaction, 10, 23
Oscillatory burning, 32, 33
Otto engine, 58
Oxygen-consumption zone, 66–68
Ozone decomposition flame, 45–47

Packing fraction, 79
Partial-equilibrium approximation, 44, 51
Peclet number, 11
Peters scale, 125
Pollutants, 59, 132, 153
Polyhedral flames, 38
Prandtl number, 10, 11
Preferential diffusion, 38
Preheat zone, 24–26, 46, 47, 53
Premixed flames, 15, 16, 21–55, 119–32
Pressure, 8–12
Probability-density functionals, 146
Probability-density functions, 135–37, 140–42, 146, 147
Pyrolysis, 68, 69, 75, 79

Quenching diameters, 98
Quenching distances, 92, 98

Radiant heat-flux vector, 9, 10, 12
Radical pool, 84
Radicals in flames, 40, 42, 52, 53, 65–69
Ramjet, 118, 131
Rankine-Hugoniot relations, 17
Rayleigh line, 17, 18, 100, 101
Reactant depletion, 86–89, 93
Reactant leakage, 66–68, 71–73

Reaction mechanisms, 39–53, 65–69, 105
Reaction order, 10
Reaction rate, 10, 23
Reaction-rate constant, 8–10, 40, 41
Reaction-sheet regime, 115–28, 133
Reaction-zone structure, 46–53, 64–68, 94, 99
Reaction-zone thickness, 24–28, 65, 66, 99, 102
Reactive-diffusive zone, 24–26, 30, 64, 65, 94
Recirculation, 131
Recombination, 45, 46, 48, 51
Reduced mechanisms, 49–55, 66, 67
Renormalization, 127, 129, 130
Reynolds number, 10, 11, 14, 113, 116, 118
Reynolds transport, 143, 144
Rocket motors, 5, 58, 155
RRA, 15, 43–45, 52–54, 67, 73

Scalar dissipation, 137, 139, 143, 144
Schmidt number, 10, 15
Shale oil, 154
Shear stress, 9, 12
Shock tubes, 86, 101
Shock waves, 8, 91, 100–102
Smoked-foil method, 104
Solid propellants, 15, 58
Soot formation, 6, 68, 69
Spark ignition, 5, 21, 58, 118
Species conservation equation, 10–13
Specific heat, 8–12, 22
Specific reaction-rate constant, 40–41, 50
Spinning deflagration, 37, 38
Spinning detonation, 104, 105
Spontaneous ignition, 90–93
Spray combustion, 58–61
Spray equation, 60
Stagnation-flow burner, 21
Standing detonation, 101
Steady-state approximation, 43–51
Stefan-Maxwell equations, 12, 13
Stirred reactor, 118, 119
Stoichiometric coefficients, 10
Strain, 71–74, 121–26, 139

Strain rate, 71–74, 122–24
Strange attractor, 147
Stratified charge, 58
Streamlines, 34
Stress tensor, 12
Sulfur oxides, 153
Supersonic combustion, 14, 58, 118, 132, 155
Surface of fuel involvement, 79
Surface-to-volume ratio, 80
Swirl, 122

Taylor scale, 117–18, 144
Temperature, 7–9
Thermal conductivity, 8, 9, 23
Thermal diffusion, 8, 9, 13
Thermal diffusivity, 26, 79, 80
Thermal explosions, 85–90, 95, 96
Thermal inertia, 80
Thermal responsitivity, 80
Thermal runaway, 83–85
Thermal thickness, 80
Transfer number, 76, 77
Transport coefficients, 7–13
Triple flame, 140, 157
Triple-shock interactions, 104
Turbojet, 118
Turbulence, 113–19
Turbulent burning velocity, 120, 128–30
Turbulent diffusivity, 143, 144
Turbulent flame speed, 120, 128–30
Turbulent flames, 119–40
Turbulent kinetic energy, 113, 114, 117, 131
Two-phase flow, 58–61

Variance, 136
Velocity, 8, 9
Viscosity, 8, 9, 12
Viscous stress tensor, 12
Von Neumann spike, 100, 101
Vortex, 124, 149
Vorticity, 122, 131

Water-gas shift, 51, 53
Well-stirred reactor, 118, 119
"Will-o'-the-wisp," 33
Wrinkled laminar flames 115–17, 119–31

Zel'dovich number, 15, 24, 29, 30,
    69, 70

Zel'dovich mechanism, 136
ZND detonation structure, 101, 105